PLASTIC TOYS

Dimestore Dreams of the 40s & 50s

Bill Hanlon

Photographs by David Belda

77 Lower Valley Road, Atglen, PA 19310

DEDICATION

To Mom, Dad and Lesley

ACKNOWLEDGEMENTS

The very thought of a book devoted entirely to plastic toys of the 1940s and 1950s would have been beyond the wildest of dreams for a small group of collectors who began corresponding back in 1978. Yet despite the fact that there has never before been a serious attempt to chronicle these fascinating toys, the number of plastic toy collectors has steadily increased over the years.

At last, we have the book we have all been waiting for, thanks to Schiffer Publishing Ltd., and to the following individuals who contributed their knowledge, enthusiasm and encouragement throughout the process: Nick Argento, David Belda, Robert S. Bergman, Frank Berlin, Christopher Byrne, Shana Byrne, Bud Danien, Chuck Donovan, Harold Frutchey, T.W. Groves, Benjamin Hirsch, Charles R. Kugel, Randy Lee, William M. Lester, Frank McCormick, Robert Neal, PhD., Richard O'Brien, Jim Radican, Edward W. Rowan, Herb Samuels, Terry Sells, Dr. Islyn Thomas, O.B.E., Lionel A. Weintraub and Dinorah Wilcox. Without their help, and without the patience and understanding of my wife Gail and my children Robin nand David, this book would still be a dream.

Title page:

Coupe with Light, 4 1/2" x 1 5/8" x 1 1/2", when one AA battery is installed, bulb in center of grill lights up, blue, plastic wheels, tires marked EVER READY, underside marked, "Made in England" and "FORD," late 1940s.

Jet Hot Rod, 5 1/2" x 3" x 2 1/4", with attached driver, assorted metallic colors, plastic wheels, Thomas Manufacturing Corp., USA (No. 289), 1954-1955.

Designed by Bonnie Hensley

Copyright © 1993 by Bill Hanlon
Library of Congress Catalog Number: 93-85217

All rights reserved. No part of this work may be reproduced or used in any forms or by any means – graphic, electronic or mechanical, including photocopying or information storage and retrieval systems – without written permission from the copyright holder.

Printed in the United States of America.
ISBN: 0-88740-460-X

We are interested in hearing from authors with book ideas on related topics.

Published by Schiffer Publishing Ltd.
77 Lower Valley Road
Atglen, PA 19310
Please write for a free catalog.
This book may be purchased from the publisher.
Please include $2.95 postage.
Try your bookstore first.

CONTENTS

Introduction	4
Chapter 1 A PLASTIC PRIMER	5
Chapter 2 THE SELLING OF PLASTICS	10
Chapter 3 VALUE, CARE AND REPAIR	18
Chapter 4 RESEARCH AND IDENTIFICATION	21
Chapter 5 DR. ISLYN THOMAS O.B.E.	26
Chapter 6 AUTOMOBILES AND AMBULANCES	30
Chapter 7 JEEPS	70
Chapter 8 RACE CARS, HOT RODS AND MOTORCYCLES	76
Chapter 9 TRUCKS, CONSTRUCTION VEHICLES AND BUSES	92
Chapter 10 BOATS	143
Chapter 11 AIRPLANES AND HELICOPTERS	171
Chapter 12 SPACE TOYS	193
Chapter 13 MILITARY TOYS	210
Chapter 14 TRAINS AND TROLLEYS	230
Chapter 15 DOLLHOUSE FURNITURE	235
Chapter 16 POTPOURRI	264
Bibliography	287
Price Guide	288
Manufactures and Distributors	293
Index	294

Motorcycle with Sidecar, 4" x 2" x 1 3/4", with handlebars that steer front wheel, red and yellow, plastic wheels, Reliable Plastics Co., Canada, late 1940s to early 1950s, example shown missing windshield and taillight on motorcycle.

INTRODUCTION

Some of my fondest memories of growing up in the early 1950s are those of lazy summer afternoons spent at Davenport's Variety Store with my best friend Billy Horgan, each of us with a quarter clutched tightly in our hand.

In those days there weren't any toy stores as we know them today, for toys were still considered seasonal. If you couldn't wait until the big department stores had their "toy lands" at Christmas, a trip to the local five and dime or variety store would have to do.

The owners of Davenport's lived around the corner from us and it amazed me how they were able to greet each child by name as they entered their store.

Once inside, the first thing we would notice was the smell. It was a hearty blend of penny candy, cheap perfume, roasting nuts, and goldfish and turtle water that always seemed long overdue for a good cleaning.

After a quick "Hi!" we made a beeline to the toy department located against the left wall. There nestled in these neat little bins, created by adjustable glass partitions, were the objects of our desire.

A quick once over to see what was new and then a more serious look was always the order of the day. There were toys made of plastic, metal, rubber and wood, but it was the plastic toys that always caught our eyes with their bright shiny colors and incredible detail, not to mention their more favorable price.

If we chose wisely we could each go home with a new toy and a ten cent bag of lemon drops or candy corn, all for under a quarter! Sometimes we had to settle for a nine cent bag of candy if we wanted a fifteen cent toy because once we got up to twenty-five cents we had to "cough up" another penny for tax.

At least once a month my mother and sister and I would go downtown to Woolworth's, whose toy department dwarfed Davenport's. Unlike today's toy store where toys are stacked from the floor to the ceiling, the toys at Woolworth's were lined up neatly in rows or in bins, all at eye level. Only the more expensive plastic toys had any packaging, which was fine with us because that way we knew exactly what we were getting. While mom shopped, my sister, Lesley and I would stare wide-eyed at row after row of brightly colored doll house furniture and the latest olive drab military equipment until our mother would take us gently by the arm and lead us away, hopefully with a new toy.

Today, these plastic toys that brought us so much pleasure and helped us prepare for the real world are eagerly pursued by collectors. Their history is a story of innovation, refinement and marketing, the likes of which the world had never seen. So settle back in your favorite easy chair and get ready to relive your Dime Store Dreams.

Ladder Truck, 6" x 1 3/4" x 2", with separate metal ladders, red with painted chrome trim, rubber wheels, Hubley Manufacturing Co., USA, 1949 to early 1950s.

CHAPTER 1
A PLASTIC PRIMER

As we look around the environment we live and work in, it's hard to imagine a world without plastics. Of course this was not always the case. Long before plastics became an accepted part of our lives they had to prove themselves just like all the rest of man's creations.

The plastic toys we take for granted today have evolved over a sixty year period. Today those toys manufactured between 1938 and 1958 are of a particular interest to a growing number of collectors who appreciate them not only for their workmanship and nostalgic value, but also as a three dimensional record of the times. As this is a field of collecting that will undoubtedly continue to grow, a brief history of plastics seems most appropriate.

Plastics are generally classified as either thermosetting plastics or thermoplastics according to their physical properties.

Those plastics that are set or cured by the application of heat and cannot be remelted are known as thermosetting plastics. Examples include the phenolics (1909), the ureas (1929), the melamines (1939) and the polyesters (1941).

Those plastics that can be remelted and remolded again and again are known as thermoplastics. Examples include the cellulose nitrates (1869), the cellulose acetates (1911), the cellulose acetate butyrates (1938), the vinyls (1927), the acrylics (1931), the polystyrenes (1937), the nylons (1940) and the polyethylenes (1943).

Since all of the toys in this book are molded from thermoplastics we will concentrate only on their development.

The ancient Egyptians and Chinese discovered long ago that nature produced substances that were plastic, that is, capable of being molded or modeled. Often referred to as natural plastics, these substances, such as amber, bitumen, horn, tortoise shell and shellac, were easily shaped by applying moderate amounts of pressure and heat.

By the 19th century, the demand for these natural plastics far exceeded the supply. During the 1850s there were shortages in America of many costly imported goods such as mother-of-pearl, amber, tortoise shell and ivory. In 1868, a manufacturer of billiard equipment offered a $10,000 reward to anyone who could come up with a substitute for expensive ivory billiard balls.

An American printer, John Wesley Hyatt (1837-1920), accepted the challenge and created the first successful semisynthetic plastic, cellulose nitrate.

While others, including English inventor, Alexander Parkes, had experimented with cellulose nitrate earlier, it was Hyatt who discovered the proper combination of cellulose nitrate and camphor that resulted in a tough, easily fabricated plastic.

Hyatt's formulation was made from linters, the short fibers that cling to cotton seeds, wood pulp, nitric acid and camphor. It didn't crack or warp like earlier attemps and was patented and trademarked under the name Celluloid in 1870. This trade name was chosen because Hyatt's formulation's main ingredient was cellulose, the chief substance found in the cell walls of all plant life.

Hyatt's Celluloid Manufacturing Company of New Jersey, later purchased by the Celanese Corp. of America, built its first plant in 1875. Here Celluloid was made in sheets of various thicknesses that could be cut, carved, planed or molded under pressure and heat into desirable shapes.

Dyestuffs could be mixed with the ingredients giving it color throughout and the finished product could be easily decorated with lacquers and paints.

Despite the fact that Celluloid was highly flammable, its low cost, light weight and other desirable features had manufacturers of everything from false teeth to shirt collars clamoring for this new material.

It is well known that prior to 1900 Germany dominated the world toy market and it was Germany who gave us the first Celluloid doll in 1878. Several years later, Hyatt's Celluloid Manufacturing Co. began making dolls from scrap material under special patents granted to W. B. Carpenter in 1880 and 1881.

Celluloid rattles and bathtub floating toys appeared around 1890 and these, along with dolls, continued to be made into the 1950s despite a growing controversy over their highly flammable nature.

Despite this, Celluloid film made the moving picture industry possible and gave us the first "safety glass" in automobiles.

In an effort to reduce cellulose nitrate's flammability cellulose acetate was born in 1911. In 1930 the Tennessee Eastman Corp. of Kingsport, Tennessee, began manufacturing its own cellulose acetate to provide Eastman Kodak with photographic film at a lower price than it had been paying.

The Tennessee Eastman Corp. soon became the second largest producer of cellulose acetate material behind the Celanese Corp. of America.

The Tennessee Eastman Corp. would not stop at providing cellulose acetate material for the sheets, rods and tubes from which finished articles were made. They went one step further and in 1932 developed the first cellulose acetate molding material capable of being compression molded, extrusion molded and injection molded. Tennessee Eastman called this new wonder plastic, Tenite. The Celanese Corp. soon followed with its version called Lumarith.

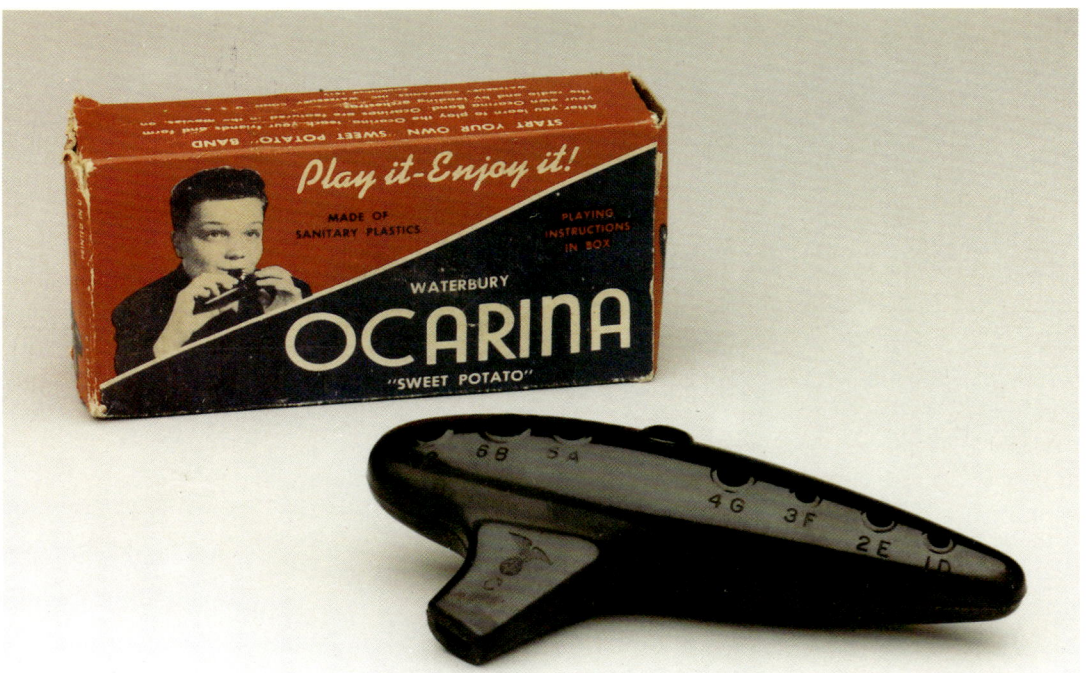

Ocarina, 6" long, compression molded using phenolic plastic, Waterbury Companies Inc., USA, 1939.

The molding press was the plastics industry's answer to mass production and once a master mold was produced in hard carbon steel, millions of identical plastic articles could be swiftly turned out with only slight finishing operations needed.

Compression molding revolutionized and dominated the plastics industry prior to the acceptance of modern injection molding in the late 1930s.

In compression molding both thermosetting plastics and thermoplastics may be used. Bakelite, Beetle and Melmac are some of the familiar trade names of thermosetting plastics that were compression molded both before and after the war. While thermoplastics can be compression molded they are usually not because of the uneconomical length of the cycle required, anywhere from fifteen seconds to fifteen minutes or longer.

In compression molding a predetermined amount of plastic powder, usually in the form of a preshaped pellet, is first loaded by hand directly into an open mold cavity. The mold is then closed and the material is squeezed into a desired shape by the application of heat and pressure. Finally, the mold is cooled and the finished piece removed. The process is somewhat like making waffles in a waffle iron.

Compression molding is ideal for making large parts like radio cabinets, steering wheels, washing machine agitators or other parts too large or too thick for injection molding.

Extrusion molding was also popular before the war. In extrusion molding only thermoplastics are used. Thermoplastic material usually in the form of granules is fed into a heated chamber where they become soft and plasticized while becoming further compressed. The material is then forced out through a die of any desired shape to form a continuous strip. This process is similar to squeezing toothpaste from a tube or dough from a cookie press.

Sheeting, tubes, rods, profile shapes and filaments, as well as coatings on wire, cable and cord are all produced in this manner.

In 1934 the first modern injection molding machines were imported from Germany. Shortly thereafter, U.S. machine-tool builders took the lead away from Germany by vastly improving their machines and the plastics industry was ready to undergo a second revolution.

This would be a revolution predicated on speed, for the high speed capabilities of the injection molding machine gave new meaning to the words, "mass production".

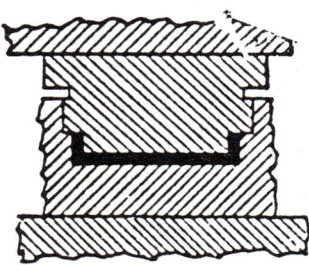

Compression molding

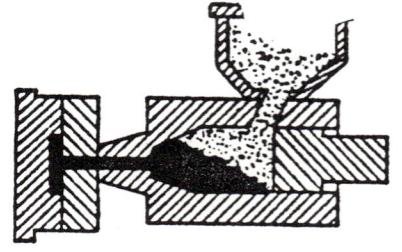

Injection molding

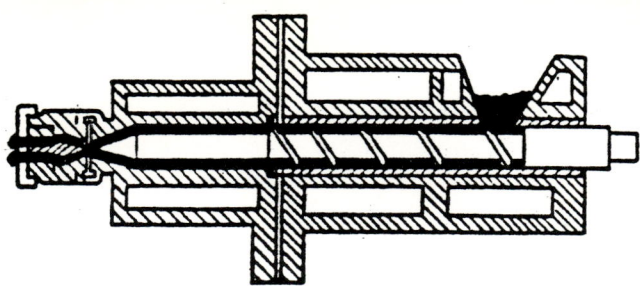

Extrusion molding

In the injection molding process, a metered charge of molding powder usually in the form of granules is gravity fed into a horizontal heated cylinder with each cycle of the machine. There it is liquified and compressed. It is then forced at high speed into a closed cold mold by a plunger or screw. The liquid plastic enters the mold by means of a "sprue" or channel which leads into the mold. Once inside, "runners" feeding off the "sprue" move the plastic to the individual cavities. There it enters the cavity through a "gate". As soon as the thermoplastic hits the cold metal, which is cooled by water, it solidifies to the desired shape. The mold then opens and the part or parts are ejected.

Spacemen, 3-3/4" tall, injection molded using polystyrene, still attached to their "sprue" by their "runners", Archer Plastics, USA, early 1950s.

Injection molding offered several advantages over compression molding. The process lent itself to complete automation and high rates of production were made possible by the short molding cycles of small parts.

Add to this, long tool and machine life, requiring a minimum of maintenance, a low ratio of mold-to-part cost in large volume production, and the fact that the left over cellulose acetate sprues and runners, along with any reject parts, could be remelted and remolded -- and you have a manufacturers dream come true.

The big break for cellulose acetate came when it was accepted by the automobile industry, which became the largest customer for plastics before the war.

Cellulose acetate knobs and other interior pieces were never unpleasantly hot or cold as sometimes happens with metal used outdoors. They could also be molded in unlimited colors to harmonize with the upholstery and other interior appointments.

Plastics were so popular with the auto industry that Tenite steering wheels compression molded on metal cores in colors to match the interior quickly replaced the standard hard rubber wheel, despite costing three times as much.

In 1942 the DuPont Co. reported that some automobiles had as many as 230 different plastic parts including interior pieces, battery boxes, and other parts that called for resistance to electricity or the action of chemicals.

Convertible, 5-1/4" x 1-3/4" x 1-1/2", injection molded cellulose acetate, possibly the first use of a clear windshield in a toy car, Irwin Corp., USA, late 1940s.

The first injection molded toys were also molded of cellulose acetate and most molders used either Tenite from the Tennessee Eastman Corp. or Lumarith from the Celanese Corp. of America.

The advantages of these toys over their metal or wood counterparts were many. Intricate shapes and fine detail which could not be incorporated in wood or metal except at excessive costs were easily attainable in plastics.

The color in plastic toys was not applied, but went all the way through. It would not chip or peel as would paint on wood or metal.

Plastic toys could incorporate transparent parts such as windshields. No other material except glass could duplicate this feat and glass was seldom used in toys because of safety.

Well designed plastic toys were difficult to break and never splintered or had sharp edges like their wooden or metal counterparts.

Plastic toys were also more hygienic because they could be easily washed. This was a big plus in toys for infants and small children.

Finally, plastic toys weighed less which meant they were easier for a small child to handle and cost less to ship, the latter contributing to a lower retail price.

If all this sounds too good to be true, it was, for these first injection molded plastic toys had their disadvantages too. The worst of these were high water absorption, fading when exposed to sunlight and poor dimensional stability. Dimensional stability is the ability of a plastic part to retain the precise shape in which it was molded. This was an inherent problem of all toys molded of cellulose acetate, many of which were rendered useless when left out in the sun or near a source of heat such as a window, heater or radiator.

To counteract these problems, some molders tried molding their toys out of cellulose acetate butyrate which offered greater moisture and sunlight resistance along with improved dimensional stability. The downside was its premium price and strong odor.

Cellulose acetate butyrate toys were molded before the war and in the immediate years following the war.

Today, toys molded of cellulose acetate butyrate can still be identified by their smell, which experienced plastic collectors relish like fine wine.

Perhaps the most successful plastic toys injection molded prior to the war were musical toys. Whistles, harmonicas, ocarinas, horns and recorders all enjoyed tremendous success.

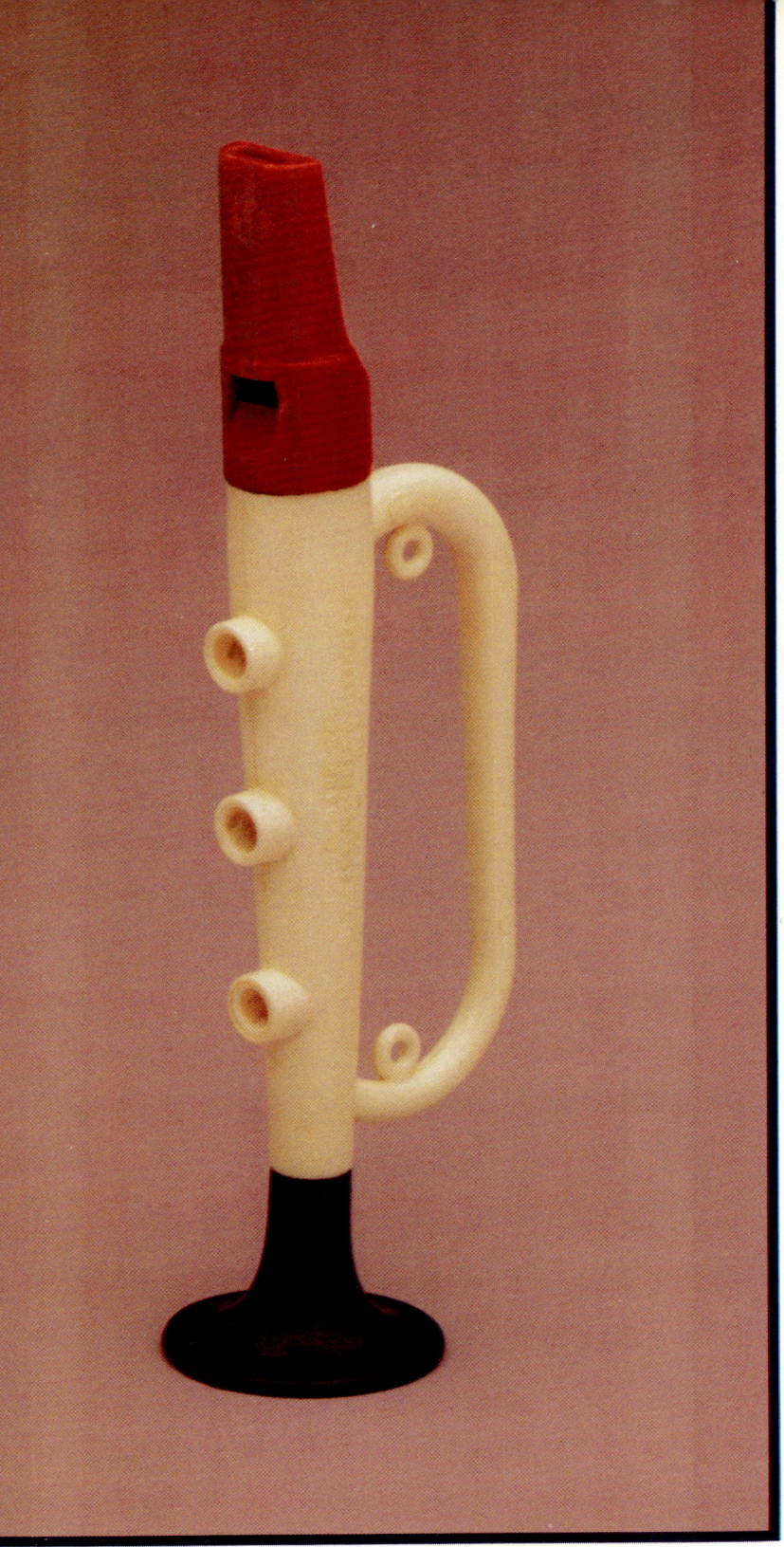

"Bugle Boy", 5-1/8" long, injection molded using cellulose acetate, Chicago Musical Instrument Co., USA, 1940.

The most successful of these was a little twenty-five cent bugle molded of cellulose acetate in red, white and blue. Offered by the Chicago Musical Instrument Co. and dubbed the "Bugle Boy", it debuted in 1940. After only 90 days the "Bugle Boy" set a new record for a plastic toy by selling over one million units. It went on to receive *Modern Plastics'* award for best plastic toy of 1940.

The next important development for the toy industry was the introduction of polystyrene, a true synthetic plastic. Polystyrene is made from ethylene (from petroleum or natural gas) and benzene (a by-product created by converting coal or coke). It was first discovered in 1839 but was not produced commercially until 1937 when the Dow Chemical Co. introduced Styron. Polystyrene possessed excellent dimensional stability, negligible moisture absorption and a greater resistance to fading than cellulose acetate.

Other advantages over cellulose acetate were its crystal-clear transparency and shiny appearance compared to cellulose acetate's yellow cast and flat appearance. The two may also be distinguished by the metallic or glass like sound polystyrene makes when struck compared to the dull or flat sound cellulose acetate makes. Polystyrene cost less than cellulose acetate and weighed less, which meant more parts per pound.

The fact is, polystyrene had everything going for it and even though some grades were very brittle and unsuitable for certain applications, the military saw its potential and became its number one user. This greatly reduced the civilian use of polystyrene until after the war.

Immediately after the war many plastics, along with metal, wood and rubber, were still in short supply. As these materials again became available for civilian use, cellulose acetate and cellulose acetate butyrate continued to be the plastics of choice for the toy industry.

Despite its lower cost, it soon became evident that the toughness of normal polystyrene was inadequate for its successful use in toys. The development of "impact resistant" polystyrenes by Dow and Monsanto eventually gave polystyrene the edge over cellulose acetate and by 1948 it was well on its way to becoming the plastic of choice for most toy applications.

By the late 1950s, toys molded of polystyrene began receiving stiff competition from toys molded of polyethylene. Polyethylene, referred to as "soft plastic" by collectors, had been around since 1943 when it first became available in commercial quantity in the United States. It offered outstanding properties including flexibility and toughness over a wide range of temperatures and a specific gravity of only 0.92 to 0.93, which was lower than any other plastic in commercial use.

While many of the companies represented in this book molded toys of polyethylene as far back as the late 1940s, the toys' rubber-like quality and lack of crisp detail have yet to catch on with collectors.

In their eyes the "hard plastic" toys, molded of cellulose acetate, cellulose acetate butyrate and polystyrene between 1938 and 1958, are the true classic plastic toys of the period.

CHAPTER 2
THE SELLING OF PLASTICS

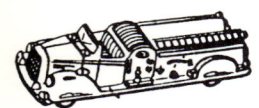

Not long after the United States entered the Second World War, industry analysts began to debate the role plastics would play in post-war America. The debate was lead by the Society of the Plastics Industry, Inc. (SPI), founded in 1937, the Society of Plastic Engineers, Inc. (SPE), founded in 1942, and the industry's leading trade journal, *Modern Plastics*, first published in 1936.

Some analysts envisioned a "Plastic Age" complete with plastic automobiles, airplanes, wearing apparel, furniture and even houses. Others simply looked forward to a renewed interest in the thousands of small plastic items that were popular before the war, but now had to be molded or fabricated out of scrap or reject material, if produced at all.

The industry had been both helped and hampered by the war. At first, the military looked upon plastics merely as a substitute for metal items such as uniform buttons, canteens, mess trays, gas mask parts, bayonet scabards, Navy dinnerware, fake weapons for training and flashlights. It wasn't long, however, until plastic became the material of choice for such items as cockpit enclosures, gun turrets, the M-74 incendiary bomb's cup and cover, the M-52 trench mortar fuse and the proximity fuse.

Military contracts for the M-52 trench mortar fuse represented the first million dollar order for a plastic part and the gold rush was on. The Ideal Novelty and Toy Co., alone, molded parts for over nine million gas masks during the war.

The proximity fuse, which cost the government about eight hundred million dollars to develop, increased the odds of rockets, missiles and bombs destroying their targets. The fuses caused the rockets and other devices to explode when they passed close to their objective, instead of on contact or by a timer.

By the war's end, components for over forty thousand fuses were being produced each day by some one hundred twenty different molders.

The military's insatiable appetite for plastics helped spread plastics technology across the country and, at the same time, dashed industry hopes that they would have enough virgin material to continue producing small goods for civilian use. When plastics were put on the list of restricted materials, most molders shifted most if not all of their business to defense work. With only scrap and reject material available for consumer items, the misuse of plastics for purposes for which they were not intended became a growing concern of the industry.

Many plastic gadgets created because of war time restrictions on metal were highly successful. Cheese graters that wouldn't rust and tea strainers that wouldn't tarnish were among the most popular items. Other plastic items that found their way onto store shelves during the war and shortly thereafter, when first rate material was still in short supply, were of dubious quality and poor design. Knives that hacked but never cut, tie-backs too flimsy to hold curtains, sink strainers that dissolved in hot water and toys that warped or were easily broken were a few of the hundreds of items that analysts feared would prejudice postwar consumers against the use of plastics for certain applications.

Pricing was another industry concern. The often spotty availability of scrap or reject plastic for desperately needed consumer goods caused prices for newly introduced items, not under price controls, to skyrocket in a non-competitive market.

The Ideal jeep, introduced in 1944, would have normally sold for ten cents, but because of the war-time market it retailed for thirty-five cents when available. Analysts argued a similar metal jeep, if available, would have sold for only fifteen cents. They warned that consumers may be willing to pay more for plastic toys at present, but after the war prices must be competitive if plastics were to capture the inexpensive toy market.

A final concern and perhaps the most controversial was educating the public. Analysts believed the public had never been educated about the differences in plastic materials and that the average consumer or salesclerk thought there was only "one" plastic.

They argued that if the irate housewife, whose son's acetate airplane warped beyond recognition when she washed it in boiling water, had been warned by a knowledgeable salesclerk or by an attached tag to use only warm soapy water, she still might be pleased with her purchase. Instead they feared she might be prejudiced against such applications for some time to come.

The question of how to educate the public and who should pay for it was the subject of numerous articles during the war and in the years immediately following it.

One group felt that molders had relied too much on the material manufacturers to promote the use of plastics and should now have their own sales organizations to promote their products.

Another group felt that if a molder chose one manufacturer's material over another's, then that manufacturer should help promote specific products made from their material.

The debate continued after the war until cellulose acetate sales took a sharp dip in 1947. Believed in part to be a consumer rebellion against misapplication, material manufacturers quickly embarked on a corrective educational program. Bulletins were sent to retail salespersons giving them precise and positive information about their merchandise. Training films and on-site seminars were made available to major chain and department stores. In

1943, Ideal Novelty and Toy Co. catalog, 8-1/2" x 11".

Toy News For the Toy Buyer, 6-3/4" x 10", Monsanto Chemical Co., USA, 1948.

Plastic Merchandiser, 8-1/2" x 11", Monsanto Chemical Co., USA, 1949.

addition, a nationwide advertising campaign was launched in major periodicals to give the general public an understanding of plastics while promoting specific products.

Of the some thirty material manufacturers who supplied about thirty-five hundred molders, laminators and fabricators, four stand out for their efforts to promote plastic toys. The Celanese Corp. of America, Dow Chemical Co., Monsanto Chemical Co. and the Tennessee Eastman Corp. all had specific programs to promote plastic toys during the late nineteen forties and early nineteen fifties.

In 1948, Monsanto mailed a ninety-six page booklet titled *Toy News for The Toy Buyer* to thousands of retail buyers across the country. Actual photos of over three hundred plastic toys molded of various Monsanto plastics were shown accompanied by a brief description, suggested retail price and the name and address of the manufacturer. The featured plastic was Monsanto's polystyrene Lustron.

In 1949, Lustron's name was changed to Lustrex which sounded a lot less like Dow's polystyrene, Styron.

That same year *Toy News for The Toy Buyer* was dropped in favor of a special issue of Monsanto's bimonthly publication *Plastic Merchandiser*, dedicated entirely to toys molded of Lustrex. Again, hundreds of plastic toys were shown, making these publications, and those that followed, indispensable in helping to identify many of the toys featured in this book, whose origins were hitherto unknown. Monsanto also ran full page four-color ads in the *Saturday Evening Post* throughout 1949 and 1950 promoting toys of Lustrex Styrene.

THE SELLING OF PLASTICS 11

Monsanto Saturday Evening Post, November 19, 1949 advertisement.

Monsanto Saturday Evening Post, July 29, 1950 advertisement.

THE SELLING OF PLASTICS 13

Styron Toys, buyer kit, 9" x 12", Dow Chemical Co., USA, 1948.

Dow had the most aggressive program of any material manufacturer for its polystyrene, Styron. Under this program, which was started in February, 1948, manufacturers were invited to submit products molded of Styron to Dow's main plant for evaluation. Products were judged on basic design, molding techniques, comparison with other plastic and non-plastic materials and their resistance to potential service hazards. Those products that passed were awarded the "Styron Label" which Dow promised to promote through a major trade and consumer advertising program.

By the end of 1948, over ten million labels were issued by Dow and by the end of 1949, 760 out of 1,900 items submitted had been rejected. Not all of the items submitted were toys but those that were and passed were given the royal treatment by Dow. The 1948 program promoting the Styron label included special Christmas displays of toys molded of Styron in 450 W.T. Grants stores and some 900 Kress and Kresge outlets. Dow provided banners, section cards, window backgrounds, and point of purchase easel displays stressing the sturdiness, color and reasonable price of Styron toys, free to any retailer responding to the campaign.

3rd Check List for Styron Toys, 8-1/2" x 11", Dow Chemical Co., USA, 1949.

THE DOW PLASTICS TOY PROMOTION
FEATURING TOYS MADE OF STYRON 475 — THE NEW, TOUGHER, SUPER-IMPACT PLASTIC

Along with the usual ads appearing in various trade-only journals, full page four-color ads promoting toys bearing the Styron label appeared in the November and December issues of the *Saturday Evening Post, Colliers, Good Housekeeping* and *Parents Magazine*. In addition to the above, Dow started mailing a series of "Check Lists for The Toy Buyer" to thousands of retailers promoting specific toys worthy of the Styron label. Like their Monsanto counterpart *Plastics Merchandiser*, hundreds of plastic toys were shown accompanied by a brief description, suggested retail price and name and address of the manufacturer.

The Celanese and Tennessee Eastman Corporations promoted toys made of their cellulose acetate plastics, Lumarith and Tenite, by running full page ads in many of the major publications already mentioned. More often than not, these ads were only one or two color and, while their programs were effective, they were far less ambitious than those of Dow and Monsanto. The reason for this was cellulose acetate toys were quickly being replaced by those made of polystyrene. In 1949 industry records show that 38,000 tons of cellulose acetate were produced compared to 100,000 tons of polystyrene.

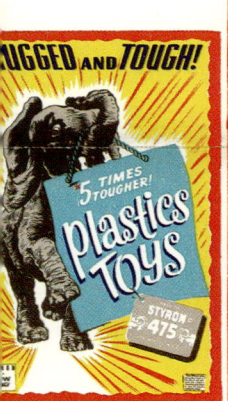
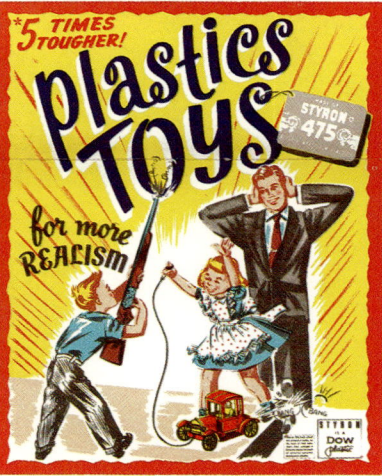

Styron ad, Dow Chemical Co., 1953. USA.

The efforts of these four major material manufacturers continued into the mid 1950s and were successful in re-introducing the toy buyer and the consumer to the colorful world of plastic toys.

From here on out, plastics would lead the toy field with clever, quick-selling toys that created greater store traffic, faster turnover and set more sales records than any of their metal counterparts.

Celanese - Ideal Playthings ad, March, 1950.

Celanese - Beton Playthings ad, April, 1950.

Celanese - Wannatoys Playthings ad, August, 1950.

Celanese - Renwal Playthings ad, September, 1950.

16 PLASTIC TOYS

Have Mom and Dad take you down to see the keen new toys IN TOYLAND!

Here's just one of the many plastic toys you must see **IN PERSON!**

"BATTLE STATIONS!"

A sea and air combat set to test your skill against your friends or family. The "target" is a red and white plastic battleship. Your choice of "weapons": plastic atom torpedoes you fire from a submarine, or plastic bombs you drop from a B-29 model plane. Hit the target and whoosh—the battleship "explodes!"

Made by Thomas Manufacturing Corp., Newark 5, N. J.

GO SEE the brand-new toys made of Monsanto plastic today. They're colorful, unusual . . . and tough! You can wash them clean in a jiffy. They won't rust . . . ever! What a smooth shiny finish—with no sharp edges to spoil your fun! They're beautifully designed, finely constructed to last a long, long time. No worry about peeling or chipping—their brilliant colors aren't just painted on—they're solid clear through. No question about it— these new plastic toys will be the most play-full toys you ever bought!
Monsanto Chemical Company, Plastics Division.

Kids! Monsanto doesn't make toys. We supply toy manufacturers with plastics from which many nifty new toys are made—materials like Lustrex styrene, Monsanto Polyethylene, and Opalon vinyl. So if you want to know more about these toys, go see them in your favorite toy store.

MONSANTO

Lustrex ad, Monsanto Chemical Co., 1953, U.S.A.

CHAPTER 3
VALUE, CARE & REPAIR

Most collectors start collecting plastic toys for their nostalgic value or because of an interest in a particular category such as trucks, motorcycles or doll house furniture. The fact that one's collection may go up in value should be looked upon solely as a secondary benefit of collecting, for the antique toy market is as fickle as the stock market. Even if your toys are never worth more than you paid for them, the very least you have done is preserved a bit of history to pass on while surrounding yourself with something that gives you pleasure. Who could ask for more?

Due to the sheer volume of hard plastic toys produced during the 1940s and 1950s, the inclusion or exclusion of a particular toy in this book makes it neither rare nor common. I am, however, convinced that the majority of toys worth collecting are presented here.

In most cases, plastic toys that are broken or missing pieces should not be purchased. A broken toy has no value unless it has a piece or pieces that can be "easily" transferred to another toy that is not broken and in need of said pieces. Do not be talked into a toy missing pieces by the seller who assures you that "you can pick up the piece or pieces you need at any flea market or toy show."

It simply isn't true. While we are on the subject, it is a good idea to start a parts box for bits and pieces which may come in handy someday.

The value of a toy is based on demand and rarity, not on the age or the particular type of plastic from which it was molded. While most collectors admire the variegated or mottled look of some pre-war and early post-war plastic toys, this marbleized look does not increase the value of a toy. In most cases the look was not intentional, but was the result of using scrap plastic!

The fact that an early toy molded of cellulose acetate or cellulose acetate butyrate is "slightly" warped should not detract from its value. Few toys molded of these plastics do not exhibit some signs of warping.

Another defect you may encounter is what appears to be unsightly burn marks on an otherwise unblemished toy. When plastic collectors first started questioning these marks it was believed the marks were made by a careless assembler hot pressing the lugs that were used to join two pieces together or hold an axle. We now know that these marks are made when two dissimilar plastics come in contact for an undetermined period of time.

The major "culprit" is vinyl, which has an excessive

Variegated U.S. Army 10 Shooter Tank, 5" x 3" x 3-1/4", when pushed, tank fires ten round pellets by means of spring powered mechanism, Bell Products Co., USA, 1957.

Assorted cleaning products and supplies for plastic.

amount of plasticizer in it to make it soft and flexible. When vinyl comes in contact with a hard plastic-like polystyrene, a chemical reaction takes place; the vinyl's plasticizer actually melts the hard plastic, in an effort to make it soft and flexible too. The rigging, harnesses and figures supplied with the Ideal pirate ship, stagecoach, chuckwagon and buckboard are famous for producing these unfortunate marks. Such marks can affect the value of a toy and care should be taken not to display or store such dissimilar plastics in contact with one another for an extended period of time.

A toy that is still in its original box will naturally have a higher value than one that is not. How much higher is something that must be decided between buyer and seller. A good "rule of thumb" is to add about 30% to the value of the toy if it still has its original box. Besides adding value to a toy, an original box could tell us the name of the manufacturer, the year the toy was made (copyright date or patent number) and the number and description of any accessory pieces included. The latter is often our only clue as to whether or not a toy is complete.

Original boxes are also appreciated for their graphic design which is often representative of the period in which they were made.

A large number of original boxes are included in the accompanying photos and hopefully you will find them as charming and attractive as I have. As with any kind of advertising, the package is often nicer than the product!

Plastic toys have a distinct advantage over their metal counterparts. In most cases, they can be safely washed with warm soapy water. Remember never to use boiling water as it may discolor the plastic or return it to a less desirable shape.

When washing a toy never use cleansers as they may scratch the surface and/or remove any hot stamped or painted details.

If the toy you are cleaning has metal axles be sure and dry them thoroughly after cleaning so they won't rust. Likewise, if your toy has a decal, friction or wind-up motor never submerge it in water or under the faucet. Instead, use clean damp paper towels or rags to clean around these vulnerable parts.

Most plastic toys weren't painted so the subject of repainting should not be an issue. Those few that were painted should not be repainted or "touched-up".

It is possible to "buff-out", by hand, small scratches on the surface of a plastic toy by using any number of plastic or automotive polishes on the market. Do not rub too hard as you may crack the plastic or damage a fragile piece next to the area you are buffing.

My favorite polish is a magnesium or aluminum wheel polish available at most automotive supply houses. With a little elbow grease you can get a "jewel-like" shine but watch out for that unsightly residue that gathers in the "nooks and cranies". This can usually be removed with cotton swabs and round wooden toothpicks.

Toys that have been vacuum metalized should never be polished as it will ruin their finish. Vacuum metalizing is a process employed to deposit a metal film on plastic. It was popular during the early 1950s and was used to increase a toy's perceived value while allowing the use of scrap or mixed-color base material which was hidden under the chrome-like finish.

Toys to be vacuum metalized are first coated with

Vacuum Metalized General Patton Tank, 6" x 3-1/2" x 3", Ideal Toy Corp., USA (No. 4944), 1953-1955.

lacquer to cover minor scratches, minimize outgassing and allow good adhesion of the aluminum. They are then placed in a vacuum chamber along with aluminum staples which are hung on tungsten filaments between two electrodes. The pressure in the chamber is reduced to .05 microns and the filaments heated to incandescence at 1200 degrees Fahrenheit. The aluminum melts and spreads evenly over the tungsten filaments until the temperature is raised to 1800 degrees Fahrenheit and the aluminum evaporates. Evaporation takes only five to ten seconds and all parts exposed receive a thin deposit of aluminum. The parts are then removed and given a top coat of lacquer to protect the metalized surface.

The finished toy was impressive looking but the aluminum film was only 1/300,000th of an inch thick and highly susceptible to wear. The good news was that one pound of aluminum would cover approximately 25,000 square feet!

Most collectors prefer to have their toys out on display for all to enjoy instead of being stored in some dark closet or attic. If you are among the majority, be careful not to display or store your plastic toys directly in front of, or close to, a window that gets direct sunlight as most plastics are heat and light sensitive. Cellulose acetate is especially prone to fading and warping and should be treated as such.

It is also not a good idea to store your plastic toys in temperature extremes such as an attic that gets over 115 degrees Fahrenheit or a basement that freezes.

With a little common sense, your plastic toys will last indefinitely and you and future generations will be able to enjoy them for years to come.

CHAPTER 4
RESEARCH AND IDENTIFICATION

The toys that follow represent a collection I started in 1978. At that time, most antique toy collectors and dealers frowned upon anything made out of plastic unless it was a model train or a 1/25 scale promotional vehicle. If a plastic toy did show up at an antique toy show, it was usually found under the table in a "junk box".

Times have certainly changed. The growing number of plastic collectors who appreciate these colorful toys for their beauty and workmanship will hopefully find new treasures to search for after reading this book.

I have tried to group similar toys into categories or chapters which should allow you to quickly find a particular type of toy. Once categorized, the toys appear in alphabetical order by manufacturer, then chronological order by date manufactured. Toys whose manufacturer is unknown are placed at the end of each chapter.

In a few instances, a toy qualified for more than one chapter. An example of this is the jeep. Jeeps have their own chapter plus they can be found in the "Military Toys" chapter and along with Roy Roger toys in the "Plastic Potpourri" chapter. Be sure to check the index for all possible locations.

I have tried to identify each toy as to its size, manufacturer's product number and the years available.

All sizes are approximate and since some have been taken from catalog descriptions, please remember it is not uncommon for a manufacturer to claim a toy is ten inches long when it is in fact only nine and three quarters inches.

Most of the toys presented have a manufacturer's name or molder's mark (logo) embossed somewhere on their underside.

A smaller group of toys molded in the years immediately following the war bear no identification whatsoever. It has been suggested that the manufacturers of these toys, many of which were of questionable quality, were simply trying to get rich quick by cashing in on the unprecedented demand for inexpensive toys and didn't care about their reputation.

A final group of toys bear the name or mark of a company that is totally misleading as to the true origins of the toy. Those toys manufactured by the Thomas Mfg. Corp. during its first five years of existence and those manufactured by the Precision Plastics Co. are perfect examples of toys whose true origins have baffled collectors for years.

The Thomas Mfg. Corp., under the direction of Islyn Thomas, designed and molded toys from 1945 until 1960. From 1945 until 1950, the Acme Plastics Mfg. Co., which was part of the Thomas Mfg. Corp. with nothing more than an office in Manhattan, did the marketing.

Toys molded from 1945 to 1950 usually bear only the Acme name, although some vehicles also have a TMC on

1948 Thomas Mfg. Corp. catalog.

their license plate. This has lead collectors to believe that Acme and Thomas were two companies when they were really only one.

When Islyn Thomas bought out Acme's Benjamin Shapiro in 1950, the Acme name was removed or supplemented with the Thomas Toy name. Toys introduced after 1950 only carry the Thomas Toy name.

Acme Plastic Helicopter, box 3-5/8" x 3-5/8" x 2", Thomas Mfg. Corp., USA, 1945-1951.

Motor Bike, 5" long, Sedan, 5" long, Station Wagon, 7" long, Convertible, 7" long, Precision Plastics Co., USA, 1947 to early 1950s.

An even more confusing tale revolves around the toys molded by the Precision Plastics Co. of Philadelphia, Pennsylvania.

Edward W. Danien, who got his start in plastics by selling custom molding for the Pyro Plastics Corp., founded the Precision Plastics Corp. in 1941. Like most other custom molders, Precision concentrated on defense work during the war years and didn't enter the toy market until after the war. In 1947 Precision introduced its first toy, a "streamlined" woodie station wagon called the "Road King". Molded of "shatter-proof" cellulose acetate from Koppers Company Inc. of Pittsburgh, Pennsylvania, it was unmarked other than the name "Triumph" which was embossed on its underside. Even examples found in their original boxes gave collectors no clue to the manufacturer. The boxes simply said "ANOTHER TRIUMPH by Triumph."

In 1948 the station wagon was joined by a sedan, convertible and motor bike. All four vehicles were only offered with a "powerful" wind-up motor and attached nickel-plated key. The previous year the station wagon was only offered as a push toy.

The convertible and motor bike carried a new "Plastic Masters" mark. The sedan was molded both with and without this new mark, and the station wagon continued to carry the "Triumph" mark.

In addition to this, all of Precision's catalogs and sales sheets bear either the U.S. Plastics Inc. name or no name at all.

If this is beginning to sound a little confusing, you can imagine my frustration in trying to identify the manufacturer of what many collectors consider to be four of the most desirable post-war plastic toys.

What finally broke the case was an article on the 1947

Toy Fair, in the June 1947 issue of *Modern Plastics*.

A photo of the station wagon was shown and credit was given to the Precision Plastics Company. As luck would have it, Precision Plastics was still in business, but now located in New Jersey. Precision's President, Bud Danien, nephew of the late Edward W. Danien, started working part time for his uncle in 1943. Bud set the record straight with this likely explanation.

Originally, Precision intended to market its toys under the "Triumph" name. In 1948, perhaps due to the overwhelming success of the station wagon, a marketing division of the company was established called U.S. Plastics Inc.. A check of an old directory shows U.S. Plastics' address as 4661 Stenton Ave. and Precision's as 4655 Stenton Avenue. Finally, the name "Triumph" was dropped in favor of "Plastic Masters" which was more descriptive of the business they wanted to promote. Case closed!

Besides applying their name or mark to the underside of a toy a few manufacturers were courteous enough to include the actual product number right there next to their name. Both Renwal and Nosco can take a bow for this helpful feature.

Unfortunately, none of the other manufacturers co-operated on this. If one does find a three or four digit number next to their name it's more likely to be the mold number rather than the actual product number. Any product number listed that differs from the number on the underside of a toy is the number that appears in catalogs or order forms and on packaging.

The Ideal Novelty and Toy Co. used a letter code, along with a number, to identify each toy and its approximate suggested retail price from 1947 to 1950. In 1951, they switched to a four digit number when they installed an IBM system.

Ideal, who had several thousand different plastic molds, usually numbered each piece with an I, followed

1949 U.S. Plastics Inc. catalog.

by a hyphen and then a three or four digit number. More than one mold number can appear on a given toy. The Ideal dump truck has mold number I-662 on the chasis and number I-818 on the bed. This tells us at least two different molds were used to produce the toy. Another mold was used to produce the wheels but they were too small to carry a number.

Sometimes a one or two digit number will appear on the underside of a toy along with the mold number or manufacturers name.

This number is the cavity number. The cavity is the depression in the mold that is filled with liquid plastic to form the toy. A mold can have one or more cavities. The more cavities, the greater the economies of production and theoretically the lower the retail price.

A mold having eight cavities compared to one, will cost the manufacturer almost eight times as much in time and money to produce and may require a larger, more costly molding machine to run it.

The cavity number appears for two reasons. The first is to let the molder know if he is having trouble with a particular cavity. If eight identical sedans are produced each cycle and all of a sudden one starts to come out with a deformed fender you can see how this number would come in handy in determining the faulty cavity.

The second reason is to discourage competitors. If one company saw that another's ten cent dump truck was "selling like hot cakes" and they wanted a piece of the action, the first thing they would do is buy a bunch of them and try to determine how many cavities the mold was to see if it was worth "tooling up" for their version.

Because of this practice many molders tried to trick the competition into thinking that the mold in question had more cavities than it really did. Thus if the highest number found on a series of identical toys was the number 10, the mold may in fact have had ten cavities but, could just as easily have had only four cavities, if they started with number seven.

The years a particular toy was manufactured is always going to be important to a collector. A toy only in production for one year may not be worth more but is certainly going to be much harder to find than one that was produced for six years. Unfortunately, the only way to tell how long a toy was offered is to have a complete run of catalogs from its manufacturer.

Since many of the manufacturers didn't offer a catalog or only offered them to the trade, it is easy to see that determining the number of years a toy was offered is a particularly arduous task.

Catalogs are often more difficult to find than the toy itself and to have a complete run of all the catalogs of a particular manufacturer is a rare occasion.

1949 U.S. Plastics Inc. catalog.

Molders' Marks

Ardee Plastics Co., Inc., Long Island City, N. Y.

Banner Plastics Corp., Bronx, N. Y.

Dillon-Beck Mfg. Co., Hillside, N. J.

Ideal Plastics Corp., Hollis, N. Y.

The Plas-Tex Corp., Los Angeles, Calif.

Pyro Plastics Corp., Union, N. J.

If we don't count the hundreds of different playsets made by Marx, then Ideal, Renwal and Thomas Toy were the number one, two and three manufacturers of hard plastic toys during the 1940s and 1950s.

In 1986 I was in the "right place at the right time" and was able to put together a complete run of Ideal catalogs from 1936 to 1986. Some of these came from the Ideal corporate offices and some came from former employees. It is believed to be the most complete collection in existence and, without it, the years the Ideal toys in this book were produced would have remained a mystery.

Over the years, fellow collectors Chuck Donovan and Terry Sells have helped provide the missing copies of the Renwal catalogs I needed to put together a run of Renwal catalogs from 1948 to 1956. These are believed the only years Renwal produced a catalog other than those for its hobby kit line.

Islyn Thomas opened up the Thomas Toy archives to give us the first in depth look at his prolific company. Thomas also provided catalogs for all but a few years of Thomas Manufacturing Corporation's existence.

It should also be noted that just because a toy isn't featured in a catalog doesn't mean a buyer from a large chain couldn't persuade a manufacturer to run a few more for them. Likewise, a small quantity of a given toy, even though not cataloged, might still be available for a time after it was discontinued.

For toys other than Ideal, Renwal and Thomas I have had to rely on incomplete runs of manufacturer catalogs, sightings in trade journals and consumer catalogs. The date of the earliest sighting is given and in most cases this method has proven to be quite accurate. However, nothing is written in stone and in a few cases a toy could have conceivably been introduced a year earlier.

I have purposely avoided trying to list every known color variation of a toy and have instead chosen to identify available colors as simply assorted colors or assorted color combinations.

Assorted colors usually means red, yellow, blue and occasionally green. Assorted color combinations might mean that the chassis of an automobile is one color, the body a second color and the interior a third color.

The type of wheel is given because the wheels may have been replaced, thus reducing the value of the toy.

I have also avoided trying to identify a given toy as a particular make or model of a real life counterpart. Most molders didn't care and it's not uncommon for a toy automobile to have the front of one make and the profile of another. An exception to the above is when a toy is listed as a particular make in a catalog. Then, I have listed it as that particular make.

Many of the toys shown came with accessory pieces and I have tried to make sure all of the pieces are shown and listed in the photo and description.

If a vehicle came with a driver, either attached or removable, I have stated so in the description. In some instances a driver, and/or other additional figures or pieces have been added to make the photos more interesting. If these are not listed in the description then they were not included with the toy.

Finally, I have included the original suggested retail price, whenever possible, to remind us all there really was a time when a dollar went further!

CHAPTER 5
DR. ISLYN THOMAS, O.B.E

If Louis Marx was the "King of Toys" then Islyn Thomas is the "Prince of Plastics", for, as you will see in the chapters that follow, no other individual contributed as much as Thomas to the development of injection molded plastic toys.

Thomas was born in Maesteg, South Wales in 1912 and immigrated with his parents to Scranton, Pennsylvania in 1923. After completing one year of high school, he attended Scranton's Johnson School of Technology, where he learned the art of tool and die making. He graduated president of his class in 1930. While furthering his education at New York University, Columbia University and the University of Scranton (where he received his doctorate in plastics engineering), Thomas began working as a tool engineer at Consolidated Molded Products Corp., also located in Scranton.

If you were interested in producing a product made entirely of plastic or incorporating plastic parts, Consolidated was "the" place to go. Formerly the Scranton Button Co., which started in 1874, Consolidated offered a wide variety of services including product development, mold design and construction, and compression and injection molding. With seven hundred and fifty employees operating on three floors, Consolidated was the largest plastics molding company in the world.

At the time, the automotive industry was the largest user of plastics, so Thomas honed his skills designing molds for numerous automotive products including dash and window crank knobs, interior trim pieces, distributor caps and hood ornaments like the famous Indian head that graced the hood of the Pontiac.

Islyn Thomas, "Prince of Plastics".

Film Strip Viewer, box 9-1/4" x 7-1/4" x 1-1/2", Acme Plastics Manufacturing Co., Inc., USA, 1940.

It was around 1935 when Benjamin Shapiro walked into Consolidated with an idea for a plastic toy filmstrip viewer. Thomas was assigned the project, and the finished product was compression molded of a phenolic plastic made by Consolidated called Arco-Lite. It was the first successful plastic toy molded or fabricated of a plastic other than Celluloid, and the start of the Acme Plastics Mfg. Co..

In 1938 Thomas became chief engineer of Consolidated. In his new position, Thomas would oversee the entire operation and be responsible for the development and molding of the necessary plastic and rubber parts that went into the Rolls Royce Merlin motors under the lend-lease program. This program was instrumental in the R.A.F.'s air victory in the Battle of Britain.

In 1941 Benjamin Shapiro wanted to expand his toy line and contracted Consolidated to start work on the molds for three toy airplanes and three toy ships. Also that year, both Consolidated and the Ideal Novelty and Toy Co. of Long Island City, New York submitted bids to the Chemical Warfare Department for a contract to injection mold outlet valves and related parts for nine million gas masks.

Ideal, the largest manufacturer of dolls and stuffed animals at the time in the United States, got its start back in 1903 when its founder Morris Michtom created the "Teddy Bear". Inspired by a cartoon about one of President Theodore Roosevelt's hunting trips, the "Teddy Bear" was an immediate success and from that point on Ideal became one of the leading innovators in the toy industry.

In 1915 Ideal introduced the first doll that could close its eyes to simulate sleep when placed on its back. In 1934 Ideal released the Shirley Temple doll, perhaps the most famous doll of all time. In 1937 Ideal made headlines again with Betsy Wetsy, the first doll that drank liquids and then needed a diaper change.

In 1940 Ideal introduced the first synthetic rubber doll with a molded plastic head, whose skin "looked and felt like human skin". By 1941 the demand for the "Magic Skin Doll" was such that Ideal's fledgling plastics division began thinking about injection molding plastic tea sets. Injection molding machines were expensive but a lucrative military contract would certainly help defray their costs and provide a market for plastic toys through the numerous Post Exchanges located throughout the country.

As fate would have it, Ideal's bid was the one accepted ... despite the fact that no one within the company had the expertise required to fulfill such a large order with strict requirements and deadlines. When Ideal couldn't deliver, a call went out to Thomas who had been responsible for Consolidated's bid and who was now both well known and respected in the industry. As Thomas tells it, "They told me to get right over to Ideal and straighten things out and that I could go over either as a civilian or in uniform! The choice was mine."

When Thomas arrived at Ideal he found twenty-four new Reed-Prentice eight ounce molding machines, most of which were sitting idle and slowly sinking into the two inch thick concrete floor which couldn't support their weight. He was immediately offered the job as General Manager of Ideal's new plastic division, and started in July of 1942.

With Thomas at the helm, not only would Ideal be assured of further government contracts but it could also set its sights on the development of plastic tea sets.

Throughout the war, only four to six of the new molding machines were used for government work; the remainder produced consumer goods. When Thomas came to Ideal, he brought along one of Consolidated's best draftsman, Harold Frutchey, who would eventually

Plastic Tea Set, box 10-1/4" x 10-1/4" x 2-1/2", includes two cups, two saucers, two knives, two forks, two spoons and two napkins, assorted colors, Ideal Novelty and Toy Co., USA, (No. TDF-59), 1943-1946. Suggested Retail $0.59.

Plastic Tea Set, box 18-3/4" x 14-1/2" x 3", includes six cups, saucers, napkins, knives, spoons, forks and plates, one tea pot, one tea pot cover, one sugar bowl, one cream pitcher; assorted colors, Ideal Novelty and Toy Co., USA (No. TSF-300), 1944-1946. Suggested Retail $3.00.

become the General Manager of the Bergen Toy and Novelty Co. in 1951. They immediately went to work on a line of plastic tea sets.

Before the advent of plastics, the ever popular toy tea set was either made of glass, which broke and was a safety hazard, or lithographed metal which dented and rusted. Now that the imported sets from Japan and Germany were no longer available and the use of metal was restricted, the category appeared doomed.

Everyone at Ideal agreed that a plastic tea set would be a tremendous success, but where were they going to get the steel to build the molds and the plastic to mold in them? New steel to make molds for consumer products and new plastic material were both on the restricted list.

Thomas again came to the rescue and was able to secure two dozen existing mold bases from his friend Ted Quarnstrom, founder of the Detroit Mold Engineering Company. A year later, Quarnstrom would develop the concept of standardized mold base assemblies which would revolutionize the whole injection molding process.

Securing the plastic was a little more involved. While at Consolidated, Thomas had worked closely with the General Shaver Division of Remington Rand to develop the case for their new electric shaver. The case was compression molded of urea plastic, a thermosetting plastic known for its wide variety of colors and electrical insulating capabilities.

In October of 1941 the Russian manufacturing city of Kharkov fell to the Germans and with it Russia's supply of pitric acid, which was essential to the war effort.

Pitric acid was also used in the synthesis of urea plastics and when Kharkov fell, the U.S. government put pitric acid on the restricted list.

1942 Ideal Victory Fleet, 8-1/2" x 11", catalog ad.

28 PLASTIC TOYS

Thomas was able to make a new mold for the razor case and injection mold it, using Styron, a polystyrene from Dow Chemical. This, however, was only until rubber was placed on the restricted list and all shaver production stopped, leaving Remmington Rand with some 800,000 pounds of Styron they had stockpiled because of expected shortages.

This was the "gold mine" of polystyrene Thomas was able to secure for Ideal, along with a good supply of clear scrap Lucite from the Firestone Tire and Rubber Company. Firestone, which had a Chemical Warfare Department contract to mold plastic lenses for gas masks, had placed a $320,000 order with Ideal for plastic tea sets. Thomas and Ideal's Abraham Katz were able to use this enormous order as leverage to secure the scrap Lucite they needed to continue producing compacts for the cosmetics industry.

To balance out Ideal's new plastics line, Thomas was able to add the molds for the three toy boats he built for Benjamin Shapiro while at Consolidated. He was also able to purchase the mold for a Bell P-39 Airacobra and an unfinished mold for a jeep from his friend Edward Rowan at the Dillon Beck Mfg. Company.

When times got tough, Thomas would buy tons of scrap cellulose nitrate and cellulose acetate from the Celanese Corp. of America. Unfortunately, these plastics were mixed together in huge piles...so the ever-innovative Thomas devised a system to separate the two. The mixture was first poured into huge drums filled with water that had salt added to raise its specific gravity to that of cellulose acetate. The cellulose nitrate, which weighed more, sank to the bottom and the desired cellulose acetate, which could be injection molded and weighed the same as the salted water, floated, and was removed.

Other military contracts at Ideal included the molding of parts for the proximity fuse, credited with helping turn the tide in the Battle of the Pacific and parts for the Manhattan Project that developed the atomic bomb.

It is of interest to note here that in 1941 a young man by the name of Lionel Weintraub started working as a general laborer in the powder room of Ideal's new plastics division. After serving in the Army, Weintraub would return to Ideal in 1946 and would eventually be elected as president of the company in 1962. Ideal would flourish under Weintraub's guidance and would establish itself as one of the premier toy companies in the United States.

In 1944 Thomas began work on the first plastic toy telephone, which was to be molded using the black scraps left over from millions of gas mask parts. Introduced by Ideal in 1945, it was one of the biggest toy hits of all time.

Islyn Thomas, who had planted the seed, would not be around to see the mighty oak grow, as he would leave Ideal to start his own company, the Thomas Manufacturing Corp., at the end of 1944. The success of Thomas Toys can best be assessed by the large number of examples that still exist today. In 1960 Thomas sold the Thomas Mfg. Corp. to the Banner Plastics Corp. of Paterson, New Jersey and became an international plastics consultant.

Thomas' accomplishments are many and he is acknowledged as one of the world's foremost experts on plastics.

In 1947 he authored the first book ever published on injection molding, titled *Injection Molding of Plastics*, which quickly became the Bible of the plastics industry.

In 1951 he served as President of the Society of Plastics Engineers, Inc.

From 1954 to 1960, in addition to being president of Thomas Mfg. Corp., he served as president of the Newark Die Co. established in 1918 and one of the foremost tool and die shops in the United States. Under his direction the Newark Die Co. would make many of the injection molds used by Louis Marx, Revell and other manufacturers of plastic toys and hobby kits during this period.

In 1975 he was invited to Buckingham Palace by Queen Elizabeth II where he was appointed an Officer of the Most Excellent Order of the British Empire for his contributions to the advancement of plastics throughout the free world.

In 1979 he was inducted into the Plastics Hall of Fame by former President Gerald Ford.

In 1987 he was elected President of the Plastics Academy and in June of 1989 was elected Chairman of the Board.

Still active today as a plastics consultant and Chairman of the Welsh American Foundation Advisory Council, Thomas was instrumental in the research for this book and his encouragement and support are greatly appreciated.

1948 Thomas Mfg. Corp. catalog.

CHAPTER 6

AUTOMOBILES AND AMBULANCES

America's love affair with the automobile is well documented and dates back to the first Model T Ford. As a toy category, it should come as no surprise that the automobile has always been among the most popular.

The first injection molded plastic toy automobiles arrived upon the scene early in 1938. They were made by the Kilgore Manufacturing Co. of Westerville, Ohio.

Our first sighting appears in the February, 1938 issue of *Playthings*, which features an ad for the Bakelite Corporation. The ad shows a toy truck, taxi, coupe, sedan and bus injection molded by the Kilgore Manufacturing Co. These five vehicles, along with an airplane, made up Kilgore's "Jewels For Playthings" line–the word "jewel" coming from the jewel-like luster of the plastic.

Kilgore had established itself as an innovator in 1932 when it introduced a line of ten-cent cast iron automobiles, with real rubber tires, challenging Tootsietoy, the leader in the inexpensive category.

Another innovative effort was recorded in the December, 1938 issue of *Modern Plastics*, which showed that Kilgore had added injection molded Tenite handles to its toy gun line, giving them that pearl-like look admired by young would-be desperadoes.

Kilgore did not produce its "Jewels For Playthings" line during the war; however, the sedan, along with the Ideal jeep, do show up in an ad for Lumarith in the 1947 issue of the *Modern Plastics Encyclopedia*. Examples of all but the bus and airplane still turn up occasionally, a testimony to the durability of cellulose acetate and the success of the Kilgore line.

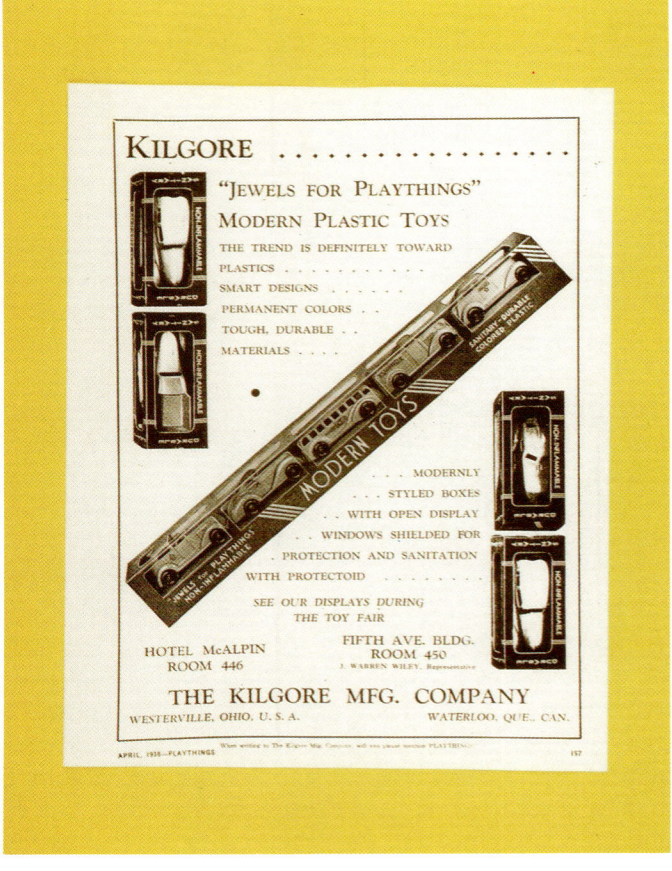

"Jewels for Playthings" ad, April, 1938, *Playthings* magazine.

"Jewels for Playthings", sedan, truck, coupe, bus, taxi, and DC-3, 3-1/2" to 4-1/2" long, Kilgore Mfg. Co., USA, 1938 to late 1940s.

Kilgore's apparent claim as the first manufacturer of an injection molded plastic toy automobile is not without challenge.

The August, 1939 issue of *Modern Plastics* magazine shows a coupe, sedan and bus manufactured by Lapin Products Inc.. They were injection molded using Tenite, a cellulose acetate from the Tennessee Eastman Corporation. One could argue that the above vehicles could have been made in 1938, but until we have proof, Lapin will have to settle for being the second manufacturer to injection mold plastic toy cars.

Lapin was established in 1912 by Max Rosenfield, a young man of twenty years. By 1929, Rosenfield had two plants producing notions and novelties made from Catalin, a colorful cast phenolic plastic.

When Rosenfield started injection molding in 1936, he was already considered a pioneer in the plastic field.

While it is not known if Lapin produced these three vehicles immediately after the war, the coupe and sedan do turn up again in the 1946 and the 1950 Sears Christmas catalogs as vehicles in "service station" playsets by the Deluxe Game Corp. of Richmond Hill, Long Island, N.Y.

Since some Lapin vehicles have been found with the name removed, perhaps the molds were sold to the Deluxe Game Corp. who used them to produce the vehicles in the above sets.

Today, examples of the Lapin coupes and sedans can still be found. Examples of the bus, which suffered from problems of dimensional stability, like its Kilgore counterpart, are quite rare.

With the exception of a pair of sedans introduced by Ideal and Irwin early in 1945, it is doubtful that any other plastic toy automobiles were molded during the war, since most available plastic scrap was turned into toy soldiers, planes, jeeps and ships.

On September 2, 1945, World War II officially ended with the surrender of Japan aboard the U.S.S. Missouri. After four years of doing without, there was a tremendous pent-up demand for toys of all kinds, especially those that were inexpensive.

There were only a few new plastic automotive surprises waiting on the toy shelves that first postwar Christmas.: a jeep made by the Thomas Mfg. Co., and Dillon Beck Mfg. Co.'s wonderful Art Deco-styled sedan with a transparent bubble top which sold one million pieces by Christmas!

Coupe, 3-1/2" long, Dillon Beck Mfg. Co., USA, 1945 to early 1950s.

Coupe and Sedan, 4" long, Lapin Products Co., USA, 1939-1950.

AUTOMOBILES AND AMBULANCES

In the years that followed, eager customers would see one innovation after another as large and small companies alike tried their hand at producing the ultimate plastic toy automobile.

In 1946, the Ideal Novelty and Toy Co. introduced the first plastic toy automobile and utility trailer combination.

In 1947, Precision Plastics Co. and Louis Marx and Co. introduced the first plastic toy station wagons. Thomas Mfg. Corp. introduced the first plastic toy convertible and the first plastic house trailer.

In 1948, Nosco introduced the first plastic toy automobiles with windup motors using plastic gears and keys in place of metal. The Renwal Mfg. Co. introduced the first plastic toy convertible with a top that went up and down and the first trunk that opened and closed. The Keystone Mfg. Co. introduced the first plastic toy automobile with a hood that opened and the first with a gas tank that could be filled with water and drained as oil under the hood.

In 1949, the Renwal Mfg. Co. introduced the first plastic toy automobile with opening doors.

In 1950, Nosco introduced the first plastic toy automobile with a visible motor with working pistons.

In 1952, the Ideal Toy Corp. introduced the first FIX-IT plastic toy automobile.

In 1953, the Ideal Toy Corp. introduced the first plastic toy automobile with working windshield wipers.

In 1954, the Ideal Toy Corp. introduced the first talking plastic toy automobile.

By the late 1950s the less expensive American made plastic toy automobiles began receiving stiff competition from Japan and Hong Kong. American manufacturers countered the onslaught with increasingly larger vehicles that showcased over twenty years of injection molding experience. These mechanical marvels and engineering wonders were expensive to manufacture and kept all but the largest American companies from competing against the imports. For the smaller manufacturer, the "Golden Age" of plastic toy automobiles was all but over.

Garage and Two Cars, garage 2-3/4" x 2-1/2" x 2", with opening door, cars 2-1/4" x 1-1/8" x 3/4", with stationary wheels, assorted colors and color combinations, All Metal Products Co., USA, early 1950s.

Sedan, 3-3/4" x 1-3/8" x 1", assorted colors, plastic wheels, Bachmann Brothers Inc., USA, late 1950s to 1960s.

Coupe, 3-1/2" x 1-1/4" x 1-1/8", Convertible, 3-1/2" x 1-1/4" x 1-1/16", assorted colors plastic wheels, B. W. Molded Plastics, USA, early 1950s.

Ambulance, 4" x 1-3/8" x 1-1/4", assorted colors, plastic wheels, Bachmann Brothers Inc., USA, 1954 to 1960s.

Sedan, 4-1/2" x 1-3/4" x 1-1/4", assorted colors including olive drab, plastic wheels, Banner Plastics Corp., USA, 1948 to early 1950s.

Station Wagon, 4-1/4" x 1-7/8" x 1-1/2", assorted colors including olive drab, plastic wheels, Banner Plastics Corp., USA, 1948 to early 1950s. Suggested Retail $0.10.

Sedan, 3-1/2" x 1-3/8" x 1", assorted colors, plastic wheels, Bachmann Brothers Inc., USA, 1954 to 1960's.

AUTOMOBILES AND AMBULANCES 33

Town and Country Car, 5" x 2" x 1-1/2", wind-up motor with attached key, assorted color combinations, plastic wheels, Banner Plastics Corp., USA, 1949 to early 1950s. Suggested Retail $0.29.

Station Wagon, 4-1/4" x 1-1/2" x 1-5/8", assorted colors, plastic wheels, Banner Plastics Corp., USA, early to mid 1950s.

Convertible, 8-3/4" x 4" x 2-3/4", wind-up motor with attached key, blue and cream, plastic wheels, Clinford Corp., USA, 1949 to early 1950s.

Coupe, 3-1/2" x 1-1/2" x 1-1/4", with clear bubble top, assorted colors, plastic wheels, Dillon Beck Mfg. Co., USA (No. C-2), 1945 to early 1950s. Suggested Retail 1945 $0.25, 1946-1950s $0.10

Jaguar, 6" x 2" x 1-3/8", assorted colors, plastic wheels, Dillon Beck Mfg. Co., USA, 1954 through mid 1950s. Suggested Retail $0.10.

MG, 4-1/2" x 1-1/2" x 1-1/4", assorted colors, plastic wheels, Dillon Beck Mfg. Co., USA, 1954 through mid 1950s. Suggested Retail $0.10.

Sedan, 9-1/2" x 3" x 3", assorted color combinations, plastic wheels, Dillon Beck Mfg. Co., USA, early 1950s.

Super Highway Fleet, card, 12-3/4" x 2-3/8" x 3", cars, 4" long, assorted color combinations, plastic wheels, Gilmark Merchandise Corp., USA, early 1950s.

AUTOMOBILES AND AMBULANCES 35

Build-A-Car Construction Set, sedan, 4" x 1-1/2" x 1-1/2", a snap together toy, no glue necessary, assorted color combinations, plastic wheels, Gilmark Merchandise Corp., USA, early 1950s.

Sedan, two and four door, both 4" x 1-1/2" x 1-1/2", with opening hoods, assorted color combinations, plastic wheels, Gilmark Merchandise Corp., USA, early 1950s.

Jaguar, 4" x 1-1/2" x 1-3/8", Convertible, 4" x 1-1/2" x 1-3/8", opening hoods, assorted color combinations, plastic wheels, Gilmark Merchandise Corp., USA, early 1950s.

36 PLASTIC TOYS

Sedan, 5-3/8" x 2" x 1-3/4", blue with painted details, plastic wheels, Irwin Corp., USA, 1945 to late 1940s. Suggested Retail $0.10.

Ambulance, 6" x 2-1/4" x 2", white with "Toyville Ambulance" in raised painted red letters on roof, plastic wheels, Hasbro Inc., USA, 1954, part of a Doctor Kit selling for $2.98.

Sedan, 4-1/2" x 1-1/2" x 1-1/4", assorted colors, plastic wheels, Ideal Novelty and Toy Co., USA (No. AU-10), 1945-1949. Suggested Retail in 1945 $0.25, 1946-1949 $0.10.

Sedan, 4" x 1-5/8" x 1-3/8", assorted colors with painted silver trim, rubber wheels, Hubley Mfg. Co., USA, early 1950s.

AUTOMOBILES AND AMBULANCES 37

Sedan, 5-1/4" x 1-1/2" x 1-1/2", Teardrop Trailer, 3-1/8" x 1-7/16" x 1-1/8", assorted colors, plastic wheels, Ideal Novelty and Toy Co., USA (No. AT-1), 1947. Suggested Retail $0.15. Trailer only (No. TR-1), 1946-1949. Suggested Retail $0.05.

Coupe, 4-3/8" x 1-1/2" x 1-1/4", (No. SC-10), 1947-1949. Suggested Retail $0.10. Square Trailer, 3-1/4" x 1-1/2" x 1-1/2", (No. TR-5), 1947-1949. Suggested Retail $0.05. Assorted colors, plastic wheels, Ideal Novelty and Toy Co., USA, trailer reissued in olive drab from 1951-1953 as part of U.S. Signal Corps Truck and Trailer (No. 4938).

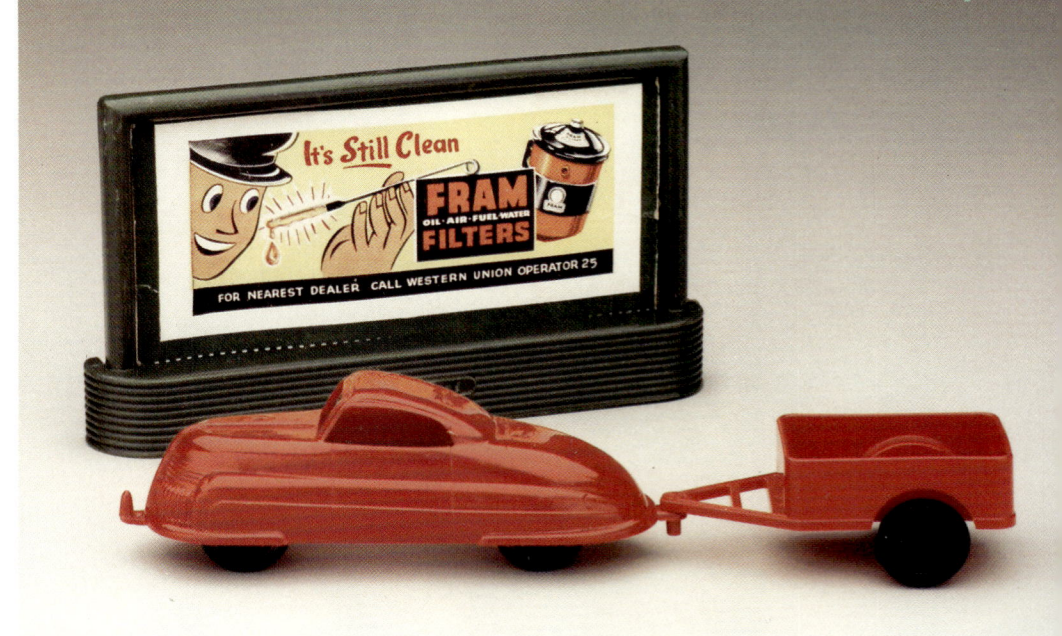

Mechanical Auto and House Trailer, sedan 5-1/4" x 1-1/2" x 1-1/2", trailer 7" x 2-1/2" x 2-1/8", wind-up motor with attached key in sedan, trailer has sliding door and detailed interior, assorted color combinations, car has rubber wheels, trailer has plastic wheels, Ideal Novelty and Toy Co., USA (No. AUMT-120), 1948-1949. Suggested Retail $1.19. Same as above, without windup (No. AHT-70), 1948-1950. Suggested Retail $0.69.

Garage, 6" x 4-7/8" x 3-5/16", with two 5-1/4" cars and/or trucks, opening overhead door, assorted color combinations, cars have plastic wheels, Ideal Novelty and Toy Co., USA (No. GAR-90), 1950, (No. 3030), 1951. (Cars vary).

Taxi, 4-3/4" x 1-7/8" x 1-7/8", yellow with red paint, plastic wheels, Ideal Novelty and Toy Co., USA (No. TA-10), 1950, (No. 3024), 1951. Suggested Retail $0.10.

Auto and House Trailer, sedan 5-1/4" x 1-1/2" x 1-1/2", trailer 7" x 2-1/2" x 2-1/8", assorted colors and color combinations, rubber wheels. In 1951, the Ideal molds were sent to Mexico resulting in this version which is the same except for rubber wheels.

AUTOMOBILES AND AMBULANCES 39

Deluxe Sedan, 10" x 4-1/32", x 3", assorted colors, rubber wheels, Ideal Novelty and Toy Co., USA (No. CA-50), 1950, (No. 3008), 1951.

Ambulance, 6-1/4" x 2" x 2-1/4", back doors open and close, white with red crosses, rubber wheels, Ideal Novelty and Toy Co., USA (No. AM-40), 1950, (No. 3004), 1951-1953. Suggested Retail $0.35.

Mechanical Police Car, 6-1/4" x 2-1/2" x 2-3/8", with friction motor and siren sound, assorted colors with decal, rubber tires on metal hubs, Ideal Novelty and Toy Co., USA (No. MPC-60F), 1950, (No. 3332), 1951-1953. Suggested Retail $0.90.

Gas Station, 8-1/4" x 6" x 2-1/2", with three 2-7/16" long cars and one 2-5/16" long tow truck, two overhead doors open, assorted color combinations, Ideal Novelty and Toy Co., USA (No. GS-100), 1950, (No. 3032), 1951-1953. Suggested Retail $1.00.

40 PLASTIC TOYS

Gas Station, box, 10" x 9-5/8" x 3-1/2", Ideal Novelty and Toy Co., USA (No. GS-100), 1950-1953.

Diner, 7" x 5-1/2" x 2-1/2", replica of roadside drive-in diner with two 2-7/16" long cars, removable transparent roof and detailed interior with separate chef, assorted color combinations, Ideal Toy Corporation, USA (No. 4254), 1951-1953.

Gas Station, box side panel, 9-5/8" x 3-1/2", Ideal Novelty and Toy Co., USA (No. GS-100), 1950-1953.

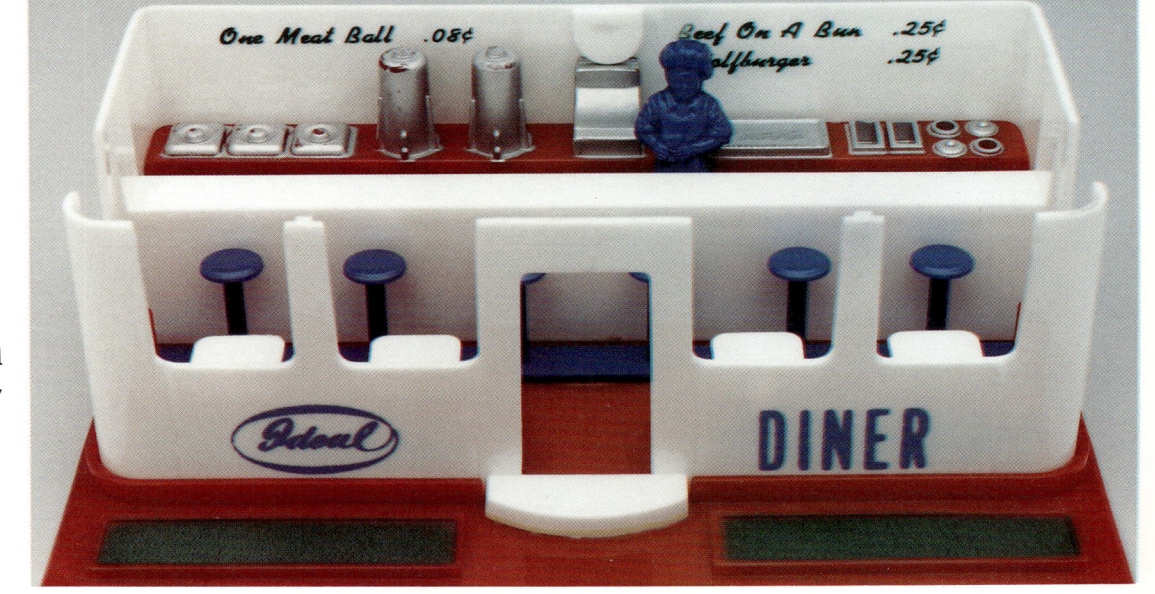

Diner, with top removed, Ideal Toy Corp., USA (No. 4254), 1951-1953.

AUTOMOBILES AND AMBULANCES 41

Car Assortment, consists of; Deluxe Sedan, 3-3/4" long, Midget Racing Car, 3-1/2" long, and U.S.A. Jeep, 3-3/8" long, assorted colors, plastic wheels, Ideal Toy Corporation, USA, 12 dozen per carton, (No. 3010), 1951-1955. Suggested Retail $0.05.

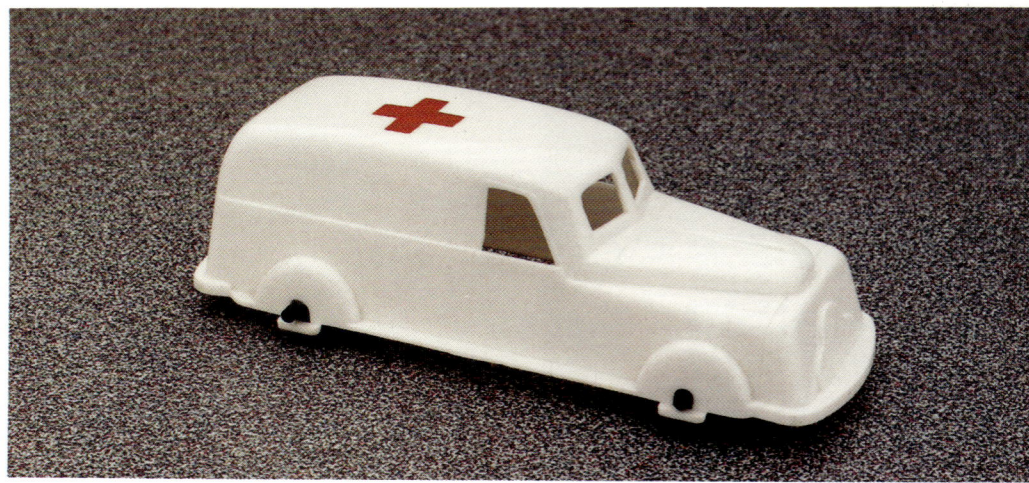

Ambulance, 4-1/2" x 2" x 1-3/4", white with red hot stamping, plastic wheels, Ideal Toy Corp., USA (No. 3066), 1952.

FIX-IT Sport Convertible, 13" x 5" x 4", hood opens to reveal motor with radiator and battery which can be filled with water, gas tank can be filled by opening cap on side, trunk opens and contains plastic working jack, gas can, spare tire, die-cast four way tire wrench, hammer, wrench, screwdriver and pry bar, all four tires can be changed, assorted color combinations, rubber tires on plastic hubs, Ideal Toy Corporation, USA (No. 3058), 1952-1955. Suggested Retail $3.00.

Car Wash, 14" x 4" x 5-1/8"; when lever is pushed back and forth, car moves forward on belt and water spurts down from nozzle; comes with U.S.A. jeep and deluxe sedan from car assortment (No. 3010), assorted color combinations, Ideal Toy Corporation, USA (No. 3031), 1952-1955. Suggested Retail $2.00.

42 PLASTIC TOYS

Luxury FIX-IT Coupe with Windshield Wipers, 10" x 3-1/4" x 3-1/2", with opening side doors, hood and trunk; all four tires can be removed by using tool kit which includes plastic working jack, screwdriver, hammer, wrench, and spare gas can, no spare tire; as car is pushed windshield wipers move from side to side; assorted color combinations, plastic wheels, Ideal Toy Corporation, USA (No. 3061), 1953-1954. Suggested Retail $1.40.

XP-600 FIX-IT Convertible, 15-1/2" x 6-1/2" x 4", opening hood and trunk; all four tires can be removed by using tool kit which includes plastic jack and emergency gas can, die-cast four-way tire wrench, hammer, screwdriver, spare tire, wrench and pry bar; car has working headlights and horn which work by means of battery; blue with white trim, rubber tires on plastic hubs, Ideal Toy Corporation, USA, (No. 3062), 1953-1955. Suggested Retail $6.00.

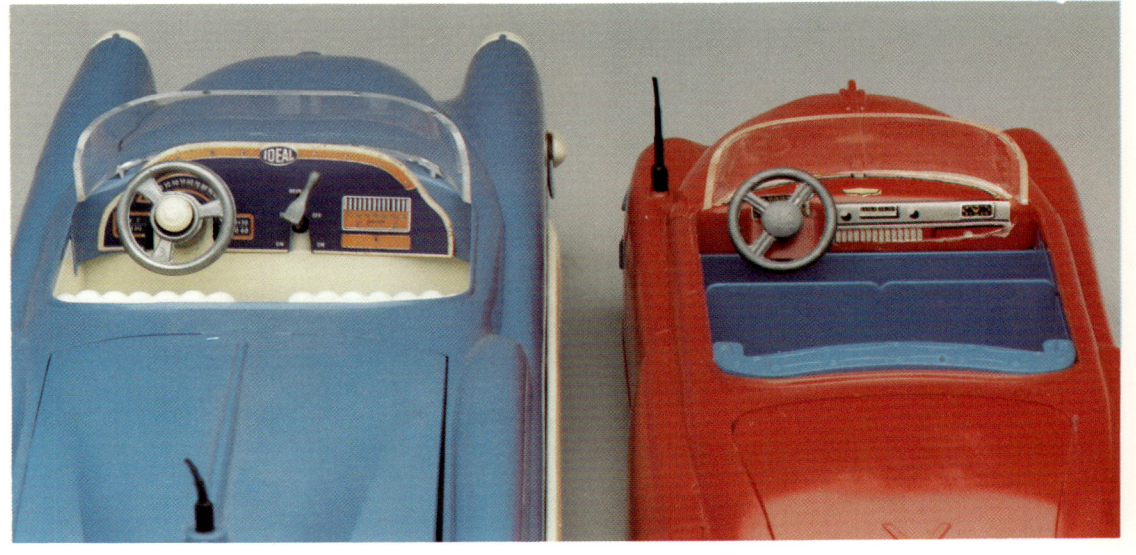

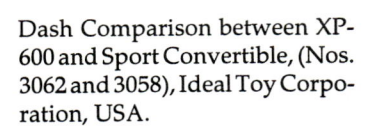

Dash Comparison between XP-600 and Sport Convertible, (Nos. 3062 and 3058), Ideal Toy Corporation, USA.

AUTOMOBILES AND AMBULANCES 43

Comic Cops Car, 6-1/4" x 3-1/2" x 2-1/4",;when pushed, front and rear seats move back and forth rocking the six 3" tall interlocking cops; assorted color combinations of yellow and black cars and three red and three blue standing or sitting cops, plastic wheels, Ideal Toy Corporation, USA (No. 4193), 1953-1954. Suggested Retail $1.00.

3 Speed MC Sport Car, 9" x 3-3/4" x 2-3/4", with hood that opens and working gear shift with three gears, forward, neutral and reverse; assorted color combinations, rubber tires on plastic spoked wheels, Ideal Toy Corporation, USA, (No. 4054), 1953-1955. Suggested Retail $1.50. Ideal would not use the true name "MG" until the above car was produced as a kit (No. 3706), 1955-1958. Suggested Retail $1.00.

3 Speed MC Sport Car, box, 10" x 4-1/2" x 3-1/4", Ideal Toy Corp., USA (No. 4054), 1953-1955.

Jaguar, 8-1/2" x 3-1/4" x 3", thirty-five detailed parts to be assembled by screws, no gluing, put it together and take it apart again and again,;opening hood, assorted colors, rubber tires on plastic "wire" wheels, Ideal Toy Corporation, USA, (No. 3082), 1954-1955. 1956-1958 as ITC Model Craft, starter set. Suggested Retail $1.00.

44 PLASTIC TOYS

Corvette, 16" x 6-3/4" x 3-3/4", sixty-one pieces to be assembled by screws, no gluing, put it together and take it apart again and again; opening hood and trunk with spare tire, removable clear plastic top, battery powered headlights, white with "Silvertone" or chrome finish trim, plastic tires on plastic hubs, Ideal Toy Corporation, USA, (No. 3083), 1954-1955. Suggested Retail $4.00.

Ferrari, 7-1/2" x 3" x 2-3/8", thirty-five detailed parts to be assembled by screws, no gluing, put it together and take it apart again and again; opening hood, assorted colors, rubber tires on plastic "wire" wheels, Ideal Toy Corporation, USA, (No. 3084), 1954-1955. 1956-1958 as ITC Model Craft, starter set. Suggested Retail $1.00.

Pegaso, 8" x 3-1/4" x 2-3/4", thirty-five detailed parts to be assembled by screws, no gluing, put it together and take it apart again and again; opening hood, assorted colors, rubber tires on plastic "wire" wheels, Ideal Toy Corporation, USA, (No. 3064), 1954-1955. 1956-1958 as ITC Model Craft, starter set. Suggested Retail $1.00.

Rolls Royce, 8-1/2" x 3" x 3", thirty-five detailed parts to be assembled by screws, no gluing, put it together and take it apart again and again; opening hood, assorted colors, rubber tires on plastic "wire" wheels, Ideal Toy Corporation, USA, (No. 3068), 1954-1955. 1956-1958 as ITC Model Craft, starter set. Suggested Retail $1.00.

AUTOMOBILES AND AMBULANCES 45

Mercedes Benz, 9" x 3-1/8" x 2-3/4", thirty-five detailed parts to be assembled by screws, no gluing, put it together and take it apart again and again; opening hood and removable roof, assorted colors, rubber tires on plastic "wire" wheels, Ideal Toy Corporation, USA (No. 3079), 1954-1955. 1956-1958 as ITC Model Craft, starter set. Suggested Retail $1.00.

Ford Thunderbird, 8-1/2" x 3-3/8" x 2-1/2", thirty-five detailed parts to be assembled by screws, no gluing, put it together and take it apart again and again; opening hood, assorted colors, rubber tires on plastic "wire" wheels, Ideal Toy Corporation, USA, (No. 3702), 1955. 1956-1958 as ITC Model Craft, starter set. Suggested Retail $1.00.

Talbot, 8" x 3-1/4" x 2-3/4", thirty-five detailed parts to be assembled by screws, no gluing, put it together and take it apart again and again; opening hood, assorted colors, rubber tires on plastic "wire" wheels, Ideal Toy Corporation, USA, (No. 3703), 1955. 1956-1958 as ITC Model Craft, starter set. Suggested Retail $1.00.

Porshe, 8-1/4" x 3-3/8" x 2-3/4", thirty-five detailed parts to be assembled by screws, no gluing, put it together and take it apart again and again; opening engine compartment, assorted colors, plastic wheels, Ideal Toy Corporation, USA, (No. 3707), 1955. 1956-1958 as ITC Model Craft, starter set. Suggested Retail $1.00.

46 PLASTIC TOYS

XP-600 Fix-It Convertible, a "stripped-down" version of (No. 3062), opening hood and trunk with plastic jack and die-cast four-way wrench to interchange tires, blue and white with bright finish front and rear bumper, vinyl tires on plastic hubs, Ideal Toy Corporation, USA, (No. 3009), 1956-1957.

FIX-IT Starfire, 20" x 7-1/4" x 6", over one hundred thirty detailed parts to be assembled by screws, no gluing, put it together and take it apart again and again; clear body, working front and rear lights, steering wheel and drive shaft work, working differential, pistons and fan belt, opening hood and trunk with spare tire, clear plastic with painted interior and trim, vinyl tires on plastic hubs, Ideal Toy Corporation, USA, 1957. Suggested Retail $14.95.

Town and Country Station Wagon, 12" x 4-1/2" x 4", windup motor with attached key, opening trunk with spare tire, roof rack with two pieces of luggage marked Irwin, blue with painted wood details, rubber wheels, Irwin Corp., USA (No. 104), late 1940s, also made as Fire Chief's Car and Taxi, both without painted wood detail.

Convertible, 5-1/4" x 1-3/4" x 1-1/2", assorted colors, plastic wheels, Irwin Corp., USA, late 1940s to early 1950s. Suggested Retail $0.15.

AUTOMOBILES AND AMBULANCES 47

Steeraway Wonder Car, 16" x 5-3/4" x 4-1/4", battery powered, with driver, Irwin Corp., USA, 1955 to late 1950s.

Sedan, 4-1/2" x 1-1/2" x 1-1/4", with driver that is suspended from the underside of the roof by a peg in his head, assorted colors, rubber wheels, Ivory, USA, early 1950s.

Sedan, 4-3/4" x 1-1/2" x 1-1/2", fill gas tank in rear with water then lift hood to drain oil, Keystone Mfg. Co., USA, 1948 to early 1950s.

Traffic Outfit, includes; 4 3/4" long sedan with opening hood and gas tank you can fill with water, two parking meters, battery powered traffic signal, and working gas pump, Keystone Mfg. Co., USA (No. 737), 1949 to early 1950s. Suggested Retail $1.98.

48 PLASTIC TOYS

Traffic Outfit, box 12-3/4" x 12-1/4" x 1-1/2", Keystone Mfg. Co., USA (No. 737), 1949 to early 1950s.

Sedan, 4-1/2" x 1-3/4" x 1-1/2", assorted colors with gold hot stamping, plastic wheels, Kilgore Manufacturing Co., USA, 1938 to late 1940s.

Taxi, 4-1/4" x 1-3/4" x 1-1/2", assorted colors with gold hot stamping, Kilgore Manufacturing Co., USA, 1938 to late 1940s.

Coupe, 4" x 1-1/2" x 1-3/8", assorted colors with gold hot stamping, plastic wheels, Kilgore Manufacturing Co., USA, 1938 to late 1940s.

Coupe, 4" x 1-3/4" x 1-1/2", assorted colors, plastic wheels, Lapin Products Co., USA, 1939-1950.

AUTOMOBILES AND AMBULANCES 49

Sedan, 4" x 1-3/4" x 1-1/2", assorted colors, plastic wheels, Lapin Products Co., USA, 1939-

Convertible, 6" x 1-5/8" x 1-1/4", assorted colors, rubber wheels, Lido Toy Corp., USA, 1949 to early 1950s.

Sedan, 5" x 1-3/4" x 1-1/2", fill gas tank in rear with water and drain under hood as oil, molded by Lido Toy Corp., USA for T. Cohen, USA, 1949 to mid 1950s.

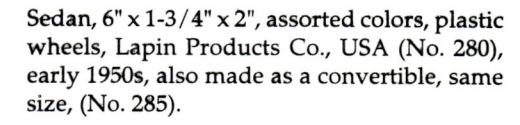

Sedan, 6" x 1-3/4" x 2", assorted colors, plastic wheels, Lapin Products Co., USA (No. 280), early 1950s, also made as a convertible, same size, (No. 285).

Sedan, (No. 287), 9" x 3" x 2-3/4", Convertible, (No. 288) 9" x 3" x 2-3/4", assorted colors, plastic wheels, Lapin Products Co., USA, early 1950s.

50 PLASTIC TOYS

Yellow Cab, 4" x 1-1/2" x 1-1/2", yellow, plastic wheels, Louis Marx and Co., USA, late 1940s to early 1950s. Suggested Retail $0.10.

Police Car, 10" x 3-3/8" x 3-3/4", with wind-up motor and attached key, green and white with decals, rubber wheels, Louis Marx and Co., USA, late 1940s to early 1950s, also made as taxi.

Yellow Cab, 4" long, Louis Marx and Co., USA, late 1940s to early 1950s.

Station Wagon and Trailer, 6" x 1-1/2" x 1-1/4", when trailer is pulled, it winds motor, assorted color combinations, rubber wheels, Louis Marx and Co., USA, 1947 to early 1950s. Suggested Retail $0.29.

AUTOMOBILES AND AMBULANCES 51

Service Station with Take-A-Part Car, station 8-1/4" x 3-1/2" x 4", sedan 4" x 1-1/2" x 1-3/8" with opening hood, assorted colors, plastic wheels, Louis Marx and Co., USA, 1953 to mid 1950s.

Traffic Light, 8" tall, battery operated, yellow, Louis Marx and Co., USA, 1950 to early 1950s. Suggested Retail $1.29.

Skyview Taxi, 5" x 2" x 1-5/8", with windup motor and attached key, yellow, rubber wheels, Louis Marx and Co., USA, late 1940s to early 1950s. Suggested Retail $0.29.

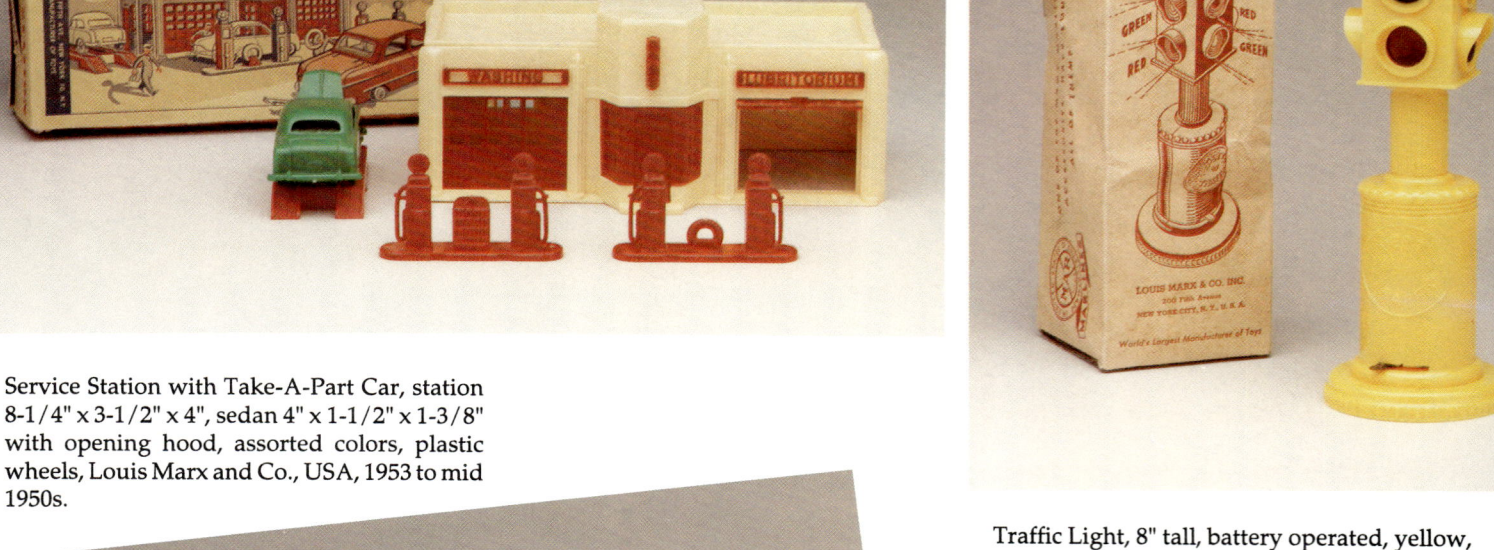

Station Wagon, 5" x 1-7/8" x 1-1/2", assorted colors, plastic wheels, Louis Marx and Co., USA, early 1950s.

Sedan, 6-1/2" x 2-1/4" x 2", assorted colors, plastic or wood wheels, Louis Marx and Co., USA, mid 1950s.

FIX-ALL Convertible, 10" x 3-1/2" x 3", take apart car with over thirty pieces, removable hard top, opening hood and trunk; includes the following tools and equipment: spare wheel and tire, gas can, fire extinguisher, adjustable jack, hammer, open end wrench, box wrench, screwdriver and adjustable wrench; assorted color combinations, plastic wheels, Louis Marx and Co., Inc., USA, 1953 to mid 1950s. Suggested Retail $1.49.

FIX-ALL Sportscar, 12-1/2" x 4-1/2" x 3-1/2", take apart sportscar with over fifty pieces, opening hood and trunk, includes the following tools and equipment; spare wheel and tire, gas can, fire extinguisher, adjustable jack, hammer, open end wrench, box wrench, screwdriver and adjustable wrench, assorted color combinations, plastic wheels, Louis Marx and Co., Inc., USA, 1953 to mid 1950s. Suggested Retail $2.98.

FIX-ALL Station Wagon, 14" x 5" x 4-3/4", take apart station wagon with over sixty-four pieces, opening hood and two piece tailgate, lithographed tin interior, visible V-8 engine with working positions, driveshaft and rear end; includes the following tools and equipment: spare wheel and tire, gas can, fire extinguisher, adjustable jack, hammer, open end wrench, box wrench, screwdriver and adjustable wrench; assorted colors with wood paneling decals, plastic wheels, Louis Marx and Co., Inc., USA, 1953 to mid 1950s. Suggested Retail $4.98.

FIX-ALL Station Wagon, motor detail, Louis Marx and Co., Inc., USA, 1953 to mid 1950s.

Sports Cars, 3-1/8" to 4" long, typical cars found in Marx service station playsets, assorted colors, metal wheels, Louis Marx and Co., Inc., USA, mid 1950s.

Sportster, 5-1/4" x 2" x 1-3/4", with friction motor, assorted colors, rubber wheels, Lupor Metal Products, Inc., USA, 1954.

Convertible, 3-1/4" x 1-1/8" x 3/4", assorted colors, rubber wheels, Manoil Co., USA, 1949 to early 1950s. Suggested Retail $0.05.

Sedan, 3-1/4" x 1-1/8" x 7/8", assorted colors, rubber wheels, Manoil Co., USA, 1949 to early 1950s. Suggested Retail $0.05.

54 PLASTIC TOYS

Station Wagon, 4" x 1-1/2" x 1-3/8", with wind-up motor and attached key, assorted colors, plastic wheels, Nosco Plastics, USA (No. 6335 S.W.), 1948 to early 1950s. Suggested Retail $0.29.

Auto-Mite Sedan, 3-1/2" x 2" x 1-1/4", with wind-up motor and attached key, assorted colors, plastic wheels, Nosco Plastics, USA (No. 6360), 1949 to early 1950s. Suggested Retail $0.19.

Auto-Mite Coupe, 3-5/8" x 1-1/2" x 1-1/8", with wind-up motor and attached key, assorted colors, plastic wheels, Nosco Plastics, USA (No. 6360), 1949 to early 1950s. Suggested Retail $0.19.

Station Wagon, 7" x 2-1/2" x 2", assorted color combinations, plastic wheels, Precision Plastics Co., USA (No. SW-69), 1947.

AUTOMOBILES AND AMBULANCES

Station Wagon, 7" x 2-1/2" x 2", with wind-up motor and attached key, assorted color combinations, rubber wheels, Precision Plastics Co., USA (No. SW-69), 1948 to early 1950s.

Sedan, 5" x 2" x 1-1/2", with wind-up motor and attached key, assorted colors, plastic wheels or rubber wheels, Precision Plastics Co., USA (No. P-250), 1948 to early 1950s.

Convertible, 7" x 2-1/4" x 1-3/4", with wind-up motor and attached key, assorted color combinations, rubber wheels, Precision Plastics Co., USA (No. R-252), 1948 to early 1950s.

Station Wagon, 5" x 2" x 1-3/4", assorted colors, plastic wheels, Premier Products Co., USA, early 1950s.

56 PLASTIC TOYS

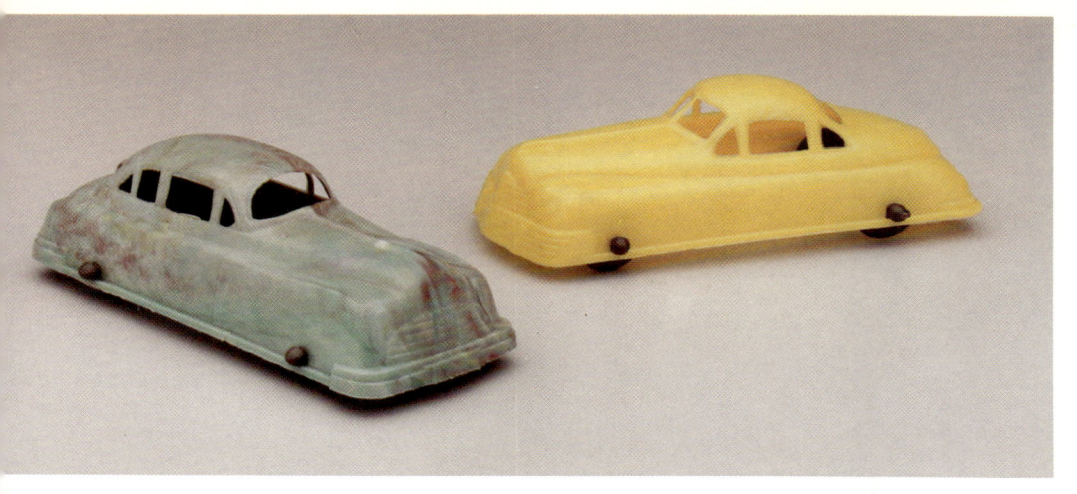

Sedan, 4-5/8" x 1-7/8" x 1-1/2", Coupe, 4-3/4" x 1-7/8" x 1-1/2", assorted colors, plastic wheels, Premier Products Co., USA, early 1950s, examples shown are polyethylene.

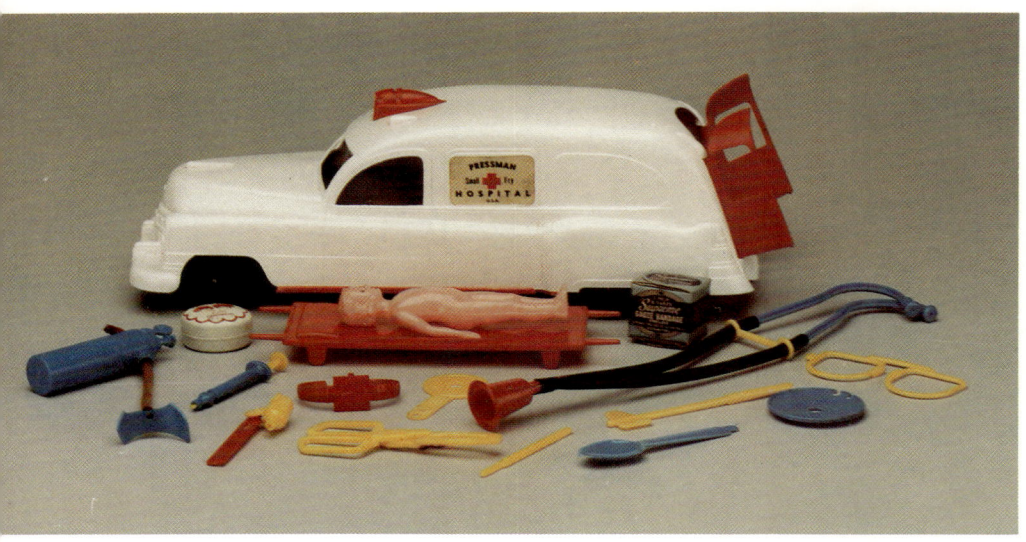

Ambulance, 14" x 5" x 4-1/2", with bell that rings when ambulance is pushed, opening rear doors, thirty medical and first aid play parts including doll patient, stretcher, watch, eye and ear examiner, hypodermic, stethoscope, surgical scissors, oxygen tank and mask, reflex hammer, thermometer, cup, head reflector, armband, spoon, eye tester, adhesive bandages, roll of gauze bandages, eyeglasses, can of burn ointment and more; white, rubber wheels, Pressman Toy Corp., USA, early 1950s.

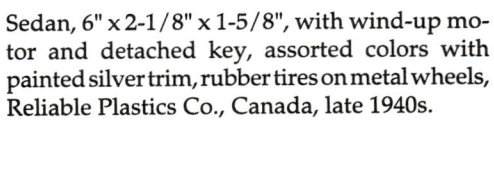

Sedan, 6" x 2-1/8" x 1-5/8", with wind-up motor and detached key, assorted colors with painted silver trim, rubber tires on metal wheels, Reliable Plastics Co., Canada, late 1940s.

Sedan, 4-1/2" x 2" x 1-1/2", assorted colors, plastic wheels, Reliable Plastics Co., Canada, late 1940s.

AUTOMOBILES AND AMBULANCES

Sedan, 4-1/4" x 1-3/8" x 1-3/8", assorted colors, plastic wheels, Renwal Manufacturing Co., Inc. USA (No. 59), 1948-1949. Suggested Retail $0.10.

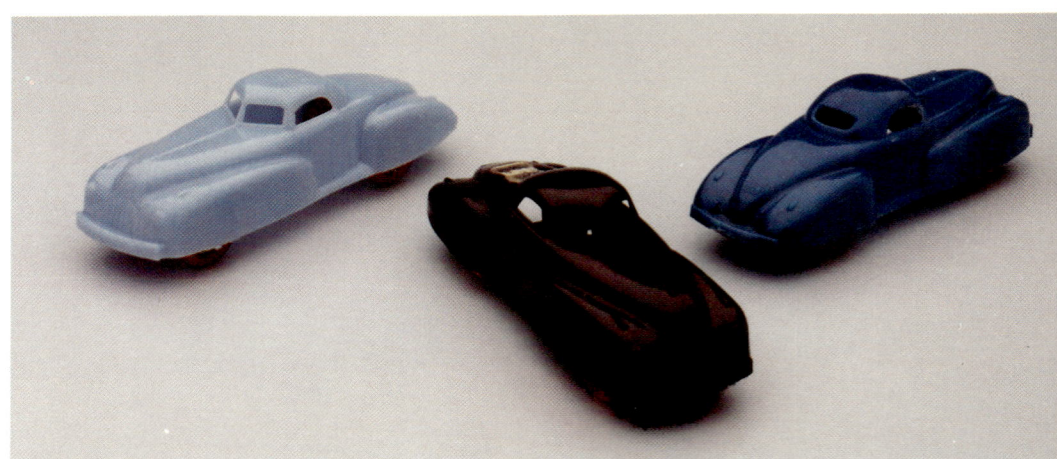

Coupe, 4-1/4" x 1-3/8" x 1-3/8", assorted colors, plastic wheels, Renwal Manufacturing Co., Inc. USA (No. 60), 1948-1949. Suggested Retail $0.10.

Convertible Sedan, with driver, top goes up and down, opening trunk with spare wheel, assorted color combinations, plastic wheels, Renwal Manufacturing Co., Inc. USA (No. 39), with painted chrome trim (No. 2039), 1948-1955. Suggested Retail $0.49.

Car and Whistle Set, five piece set includes; (No. 45) pistol whistle 2-5/8" x 1-3/4" x 1", (No. 59) sedan, (No. 60) coupe, (No. 61) racer without driver and (No. 62) truck, Renwal Manufacturing Co., Inc. USA (No. 501), 1948-1949, 1950 racer with driver replaces racer without and coupe No. 102 and sedan No. 103 replace No. 61 and No. 59.

58 PLASTIC TOYS

2-Door Sedan, 6-1/2" x 2-3/8" x 2-1/4", with driver, opening doors and trunk with spare wheel, assorted color combinations, plastic wheels, Renwal Manufacturing Co., Inc. USA (No. 90), with painted chrome trim (No. 2090), 1949-1955. Suggested Retail $0.39.

2-Door Sedan Construction Kit, 6-1/2" x 2-3/8" x 2-1/4", red, plastic wheels, Renwal Mfg. Co., Inc., USA (No. 176), 1953-1955. Suggested Retail $0.39.

Taxicab, 6-1/2" x 2-3/8" x 2-3/8", with driver, opening doors and trunk with spare wheel, assorted color combinations, plastic wheels, Renwal Manufacturing Co., Inc. USA (No. 91), with painted chrome trim (No. 2091), 1949-1955, as a construction kit (No. 177), 1953-1955. Suggested Retail $0.39.

Taxicab, 6-1/2" x 2-3/8" x 2-3/8", original packaging, Renwal Mfg. Co., Inc., USA (No. 91), 1948-1955.

AUTOMOBILES AND AMBULANCES 59

Garage with Two Cars, 6-1/2" x 5" x 3-1/2", with opening doors and two 4-1/4" long cars, assorted color combinations, plastic wheels, Renwal Manufacturing Co., Inc. USA (No. 92), 1949 garage only, 1950-1955 garage and two cars. Suggested Retail $0.98.

Coupe, 4-1/4" x 1-5/8" x 1-1/2", assorted colors, plastic wheels, Renwal Manufacturing Co., Inc. USA (No. 102), with painted chrome trim (No. 2102), with painted silver Fire Chief trim (No. 151), with painted silver or blue Police trim (No. 152), 1950-1956. Suggested Retail $0.10.

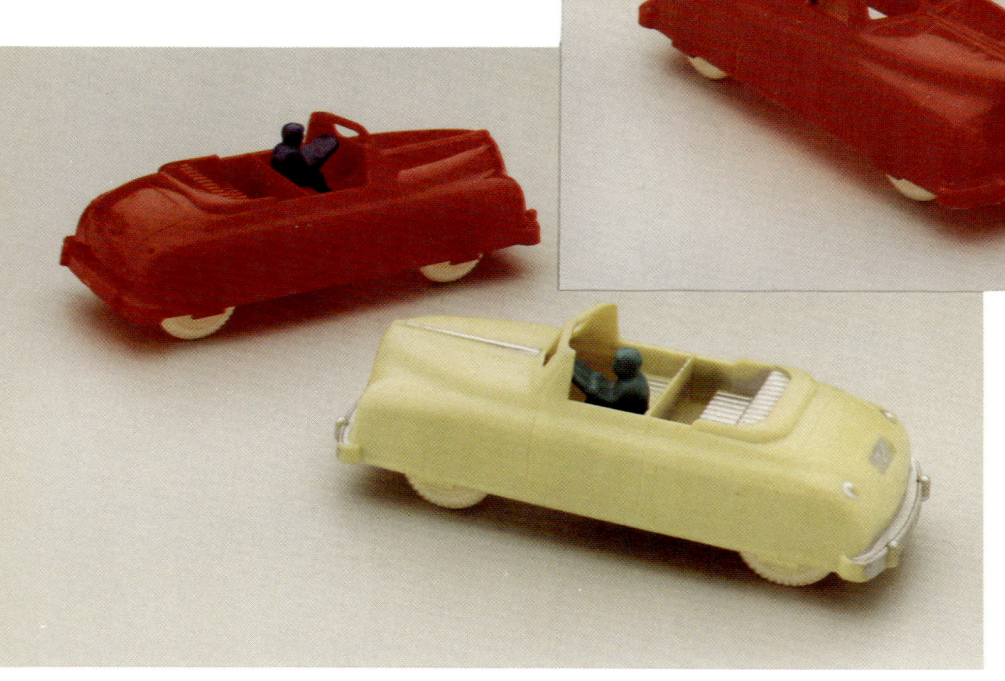

Sedan, 4-1/4" x 1-3/4" x 1-1/2", assorted colors, plastic wheels, Renwal Manufacturing Co., Inc. USA (No. 103), with painted chrome trim (No. 2103), with silver painted Taxi trim (No. 153), 1950-1956. Suggested Retail $0.10.

Convertible, 4-1/4" x 1-3/4" x 1-5/8", with driver, assorted colors, plastic wheels, Renwal Manufacturing Co., Inc. USA (No. 104), with painted chrome trim (No. 2104), 1950-1956. Suggested Retail $0.10.

Convertible, 9-1/2" x 3-1/4" x 3-1/4", with friction motor and driver, top goes up and down, doors open and close, opening trunk with spare wheel, assorted color combinations, rubber tires on plastic wheels, Renwal Mfg. Co., Inc., USA (No. 106), 1950-1952. Suggested Retail $1.29. Variation with only top that raises and lowers (No. 206), 1953-1954. Suggested Retail $0.79.

Auto Jack, 3-3/4" x 1-3/16" x 7/8", a working jack, assorted color combinations, Renwal Manufacturing Co., Inc. USA (No. 110), 1950-1953. Suggested Retail $0.10.

Sedan, 3-1/8" x 1-1/8" x 1-1/16", (No. 143), Coupe, 3-1/16" x 1-1/8" x 1-1/16", (No. 144), Convertible, 3-1/8" x 1-1/16" x 1-1/16", (No. 147), 1950-1955, assorted colors, plastic wheels, Renwal Manufacturing Co., Inc. USA. Suggested Retail $0.05 each.

Cadillac Convertible, 5-1/2" x 1-3/4" x 1-7/8", with driver, top goes up and down, assorted color combinations, plastic wheels, Renwal Manufacturing Co., Inc. USA (No. 174), 1953-1955. Suggested Retail $0.19.

AUTOMOBILES AND AMBULANCES 61

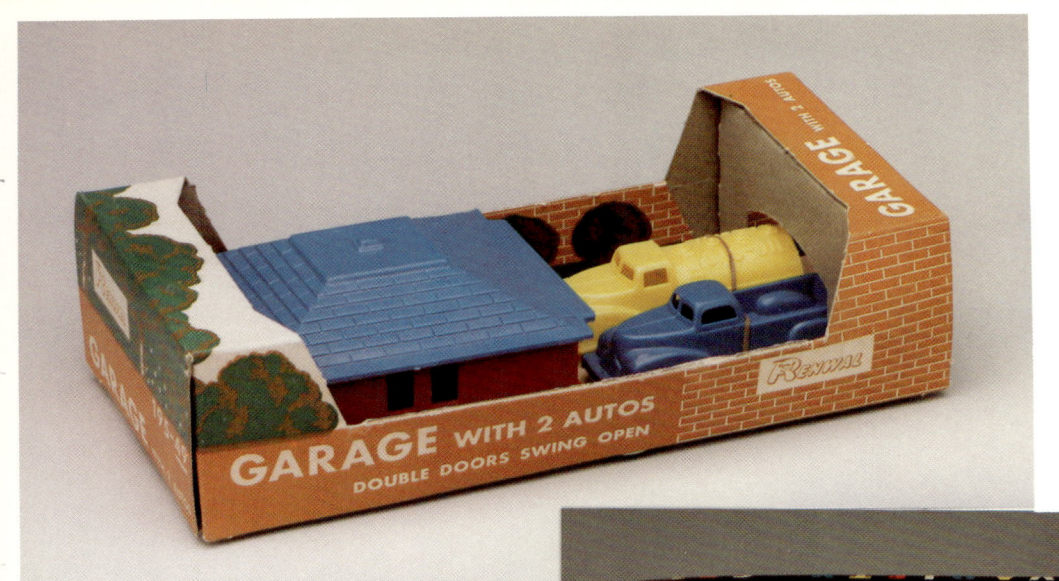

Two-Car Garage, 4-1/8" x 3-5/8" x 2-1/4", with opening doors and two 3 1/8" long automobiles, Renwal Manufacturing Co., Inc. USA (No. 195), 1953-1955. Suggested Retail $0.49.

Highway Variety Set, 4" to 4-3/8" long, includes (No. 123) school bus, (No. 124) city bus, (No. 2061) racer, (No. 2088) racer, (No. 2093) delivery truck, (No. 2094) gasoline truck, (No. 2102) coupe, (No. 2103) sedan, (No. 2104) convertible, and (No. 2621) truck, assorted colors, plastic wheels, Renwal Manufacturing Co., Inc. USA (No. 302), 1953-1954. Suggested Retail $0.98.

Old Fashioned Car, 8-1/2" x 4" x 3-3/4", with driver, assorted color combinations, plastic wheels, Renwal Manufacturing Co., Inc. USA (No. 201), 1954-1955. Suggested Retail $0.69.

Toytown Service Garage Set, 7" x 5" x 2-7/8", packaging creates a full service garage with illustrations of gas pumps, tire racks and working ramp to parking area, includes five 3-1/4" cars and trucks, assorted colors and combinations of colors, Renwal Manufacturing Co., Inc. USA (No. 218), 1955. Suggested Retail $2.00.

Convertible, 3" x 1-1/4" x 1", with woman driver, dog in rear seat, assorted colors, plastic wheels, Ross Tool and Mfg. Co., USA, 1949 to early 1950s.

Sedan, 7" x 3" x 2-1/2", red, rubber wheels, Royal Plastics Inc., USA, late 1940s to early 1950s.

Sedan, 9-1/4" x 3-3/4" x 2-1/2", red, rubber wheels, Royal Plastics Inc., USA, late 1940s to early 1950s.

Jaguar, 8-3/4" x 3-1/8" x 2-5/8", with friction motor and driver, removable hood, red, rubber tires on plastic wheels, Saunders Tool and Die Co., USA (No. 700), 1952 to mid 1950s. Suggested Retail $1.29.

AUTOMOBILES AND AMBULANCES 63

Station Wagon, 4-1/2" x 1-3/4" x 1-1/2", assorted colors, plastic wheels, Slik-Toy, USA, late 1940s to early 1950s.

Auto Pencil Sharpener, 2-3/4" x 1" x 1-1/4", top comes off to empty, assorted color combinations, Sterling Plastics Co., USA, 1950 to mid 1950s.

Convertible Coupe and House Trailer, convertible, 4-1/4" x 1-3/8" x 1-1/4", detachable trailer, 3-7/8" x 1-3/8" x 1-5/16", assorted colors, rubber wheels, Thomas Manufacturing Corp., USA (No. 30), 1947-1950.

Airline Limousine and Streamlined Utility Trailer, limousine 4-1/2" x 1-1/4" x 1-1/4", trailer 2-1/2" x 1-1/2", assorted colors, rubber wheels on limousine and plastic wheels on trailer, Thomas Manufacturing Corp., USA (No. 55), 1948-1951. Suggested Retail $0.10. Limousine (No. 29), 1947-1951, trailer (No. 48), designed to fit Numbers 27, 29, 40, and 41, 1948 to early 1950s. Suggested Retail $0.05.

64 PLASTIC TOYS

Sedan, 4-5/8" x 1-3/8" x 1-1/4", assorted colors, rubber wheels, Thomas Manufacturing Corp., USA (No. 27), 1947-1951. Suggested Retail $0.10. Vacuum metalized sedan (No. 98), 1950 to mid 1950s. Sedan, 4-5/8" x 1-3/8" x 1-1/4", fill gas tank in rear with water and drain under hood as oil, assorted colors, plastic wheels, molded by Thomas Mfg. Corp., USA, believed to be converted by Keystone Mfg. Co., USA, as a stop-gap until their similar vehicle could be molded, 1947-1948.

Sedan with Canoe, sedan 4-5/8" x 1-3/8" x 1-1/4", canoe 3-3/8" x 1-1/16" x 1/2", sedan missing plastic roof rack to hold canoe, assorted colors, rubber wheels, Thomas Mfg. Corp., USA, 1954-1955.

Police Radio Car, 4-1/2" x 1-1/2" x 2", with separate siren, spotlight and radio antenna, assorted colors with pressure sensitive decal on roof, rubber wheels, Thomas Manufacturing Corp., USA (No. 67), 1949-1951, also made as Fire Chief Radio Car, same No. 67 but with different decal.

Streamlined Sedan, 4-1/2" x 1-1/2" x 1-1/2", (No. 77S), 1949-1956, House Trailer, 5-3/8" x 1-7/8" x 1-7/8", 1952-1955, assorted colors, rubber wheels, Thomas Manufacturing Corp., USA.

AUTOMOBILES AND AMBULANCES 65

Streamlined Convertible Coupe, 4-1/2" x 1-1/2" x 1-1/2", assorted colors, rubber wheels, Thomas Manufacturing Corp., USA (No. 77C), 1949 to mid 1950s, driver added in 1950.

Taxi, 4-1/2" x 1-1/2" x 1-7/8", with separate taxi sign on top, assorted colors with two water decals - one on the side and one on top, rubber wheels, Thomas Manufacturing Corp., USA (No. 128), 1950 to mid 1950s.

Kar-Kit, 11" x 5" x 3-1/2", with or without wind-up motor, removable top, car can be assembled and disassembled by nuts and bolts, assorted color combinations, metal frame, rubber wheels, Toy Founders Inc., USA, 1947. Suggested Retail with motor $6.95, without motor $5.95.

Litemobile, 5 1/4" x 1 3/4" x 1 5/8", with a battery powered light bulb in grill that lights up when front of car is pushed down and a metal spring makes contact with the threaded part of the bulb, assorted colors, rubber wheels, Manufacturer Unknown, USA, 1949 to early 1950s.

66 PLASTIC TOYS

Sedan, 5-1/4" x 2-1/8" x 1-1/2", assorted colors, plastic wheels, Cruver Mfg. Co., USA, 1947 to late 1940s.

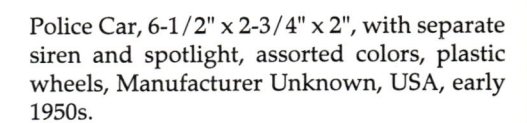

Police Car, 6-1/2" x 2-3/4" x 2", with separate siren and spotlight, assorted colors, plastic wheels, Manufacturer Unknown, USA, early 1950s.

See Through Sedan, 4-1/4" x 1-3/4" x 1-1/2", car may be taken apart and reassembled without gluing, Manufacturer Unknown, European, early 1950s.

Sedan, 9-1/2" x 4" x 3-1/4", assorted colors, rubber wheels, Manufacturer Unknown, USA, early 1950s. This car comes with a large house trailer.

AUTOMOBILES AND AMBULANCES 67

Sedan, 9-1/2" x 4" x 3", red, rubber wheels, Manufacturer Unknown, USA, early 1950s.

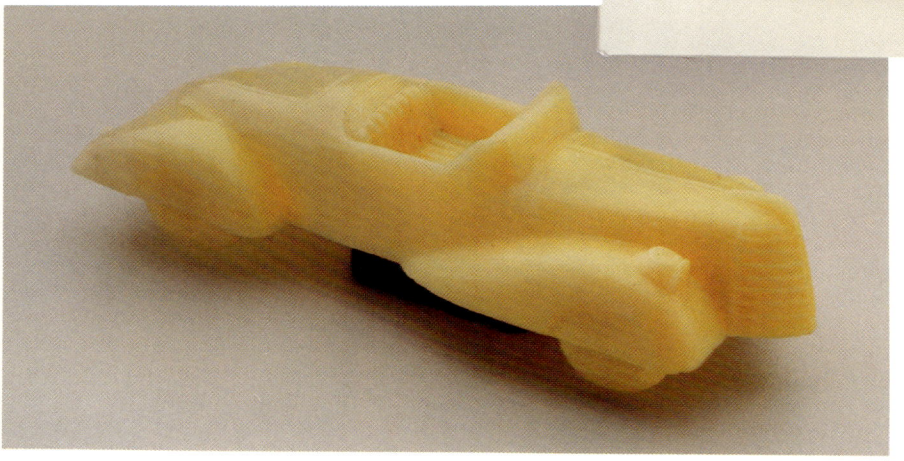

Convertible, 3" x 1" x 7/8", with attached magnet, possibly part of a driving game, yellow, stationary plastic wheels, Manufacturer Unknown, USA, late 1940s.

Sedan, 2" x 5/8" x 5/8", possible Cracker Jack prize, assorted colors, stationary plastic wheels, Manufacturer Unknown, USA, late 1940s to early 1950s.

Plasti-Kar, 6-1/4" x 2-1/8" x 2", assorted colors, plastic wheels, Tedsco Plastics Inc., USA, 1946 to early 1950s. This is the only vehicle in this book that is compression molded.

Stutz Bearcat, 5-1/8" x 2" x 1-5/8", with friction motor, red with silver painted details, rubber wheels, Louis Marx and Co., Inc., USA, mid 1950s, motor marked made in Germany; Roadster, 5-3/4" x 2-3/4" x 3-1/2", with driver and passenger, yellow, plastic wheels, Manufacturer Unknown, USA, early to mid 1950s; Touring Car, 4" x 1-1/2" x 2-1/2", with driver, red, plastic wheels, Allied Molding Corp., USA, early 1950s.

Ko-Ko the Car Clown, 7" x 5-1/4" x 7", with wind-up motor and attached key, removable clown rattle, clown goes up and down as car rolls along, assorted color combinations, rubber wheels, Saunders-Swadar Toy Co., USA, (No. 200). 1954 through mid 1950s. Suggested Retail $1.98.

AUTOMOBILES AND AMBULANCES 69

CHAPTER 7
JEEPS

No other vehicle since the Model T Ford captured as many American hearts as the jeep. Recognized by young and old alike, it somehow symbolized America's grit and determination to win the Second World War.

As America prepared itself for the inevitable, the army sent out invitations to bid on seventy light reconnaissance and command test vehicles. On July 22, 1940, the American Bantam Car Company was awarded the contract and produced the first prototype forerunner of the famous jeep in an amazing forty-nine days!

Both Willys-Overland and Ford submitted prototype vehicles later that year and they, along with Bantam, were awarded contracts for an additional fifteen hundred test vehicles each.

While there are many stories of how the little jeep got its name, one seems more believable than the rest, especially if you are a Ford man!

The Ford model was called the Ford GP which stood for General Purpose. Some believe that the model designation pronounced Gee Pee was later shortened to J-E-E-P.

The Willys-Overland model known as the GPW (General Purpose Willys) was the eventual winner and awarded a contract for eighteen thousand, six hundred jeeps on July 23, 1941. Over six hundred thousand jeeps would eventually be produced during the war years and the demand was such that Ford was chosen as the alternate production company, producing two hundred seventy-seven thousand, eight hundred and ninety-six of the above General Purpose Willys on its assembly lines.

The Dillon Beck Mfg. Co. of Hillside, New Jersey started work on the mold for the first injection molded jeep about the same time as the Quarter Master Corps awarded Willys-Overland its contract. Unfortunately, they were unable to finish the mold because of military contracts and it, along with their mold for a Bell P-39 Airacobra, was purchased late in 1942 by Islyn Thomas, General Manager of Ideal.

At Ideal considerable detail was added to the mold, and a separate windshield and a spare tire were incorporated into the design. It was added to the line in 1944, and was the only plastic jeep produced during the war. Needless to say, it was an instant hit, and demand far exceeded supply.

An article in the April, 1944 issue of *Modern Plastics* reported the Ideal jeep had been retailing for thirty-five cents, when available. In the two years immediately following the war, the suggested retail price remained the same due to the fact that many toys were still in short supply.

The Thomas Manufacturing Corp. would introduce its version of the jeep with a windshield that actually folded down late in 1945 and added a trailer for it in 1946.

The Ideal version, which sold from 1944 until 1947, was no match for the more realistic Thomas Toy version sold from 1945 well into the mid 1950s.

Another jeep, with a fold down windshield and considerably more detail than the Thomas jeep, was introduced by the Plast-Tex Corp. of Los Angeles, CA in 1948. With its thick wall sections and heavy duty metal axles, it not only looked good, but felt good! It too was available with a trailer and even a cannon to tow behind. Both of these accessories appear to have been made in limited quantities and are very rare today.

After the war the jeep was popularized in an endless stream of movies about the war and it remained a perennial favorite with outdoorsmen and the sporting set. Other toy manufacturers were quick to capitalize on the little jeep's popularity and turned out numerous versions of America's favorite vehicle.

Toy jeep sales got a big boost in 1952 when NBC introduced "The Roy Rogers Show" which included Sheriff Roy's comical sidekick, Patrick Aloysious Brady and his tempermental jeep, Nellybelle.

With one hundred and one episodes of the hit show aired between 1952 and 1957, it's hard to say whether the Second World War or Nellybelle did more for toy jeep sales!

Jeep, 4" x 1-1/2" x 1-1/2", assorted colors with decal, plastic wheels, Ideal Novelty and Toy Co., USA (No. J-1), 1944-1947.

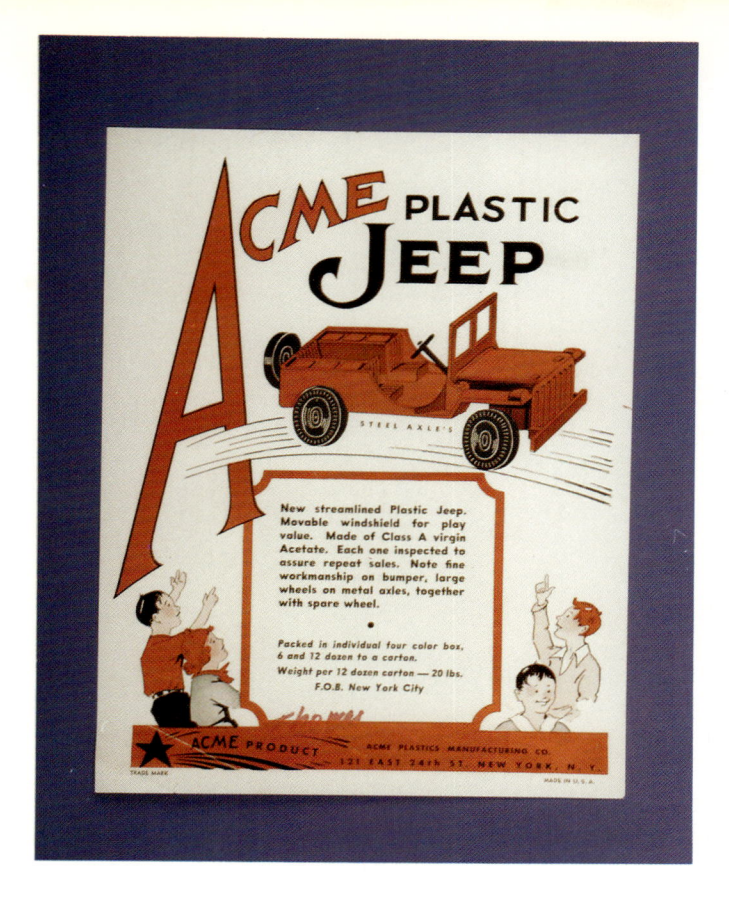

Jeep ad. Thomas Mfg. Corp. 1945.

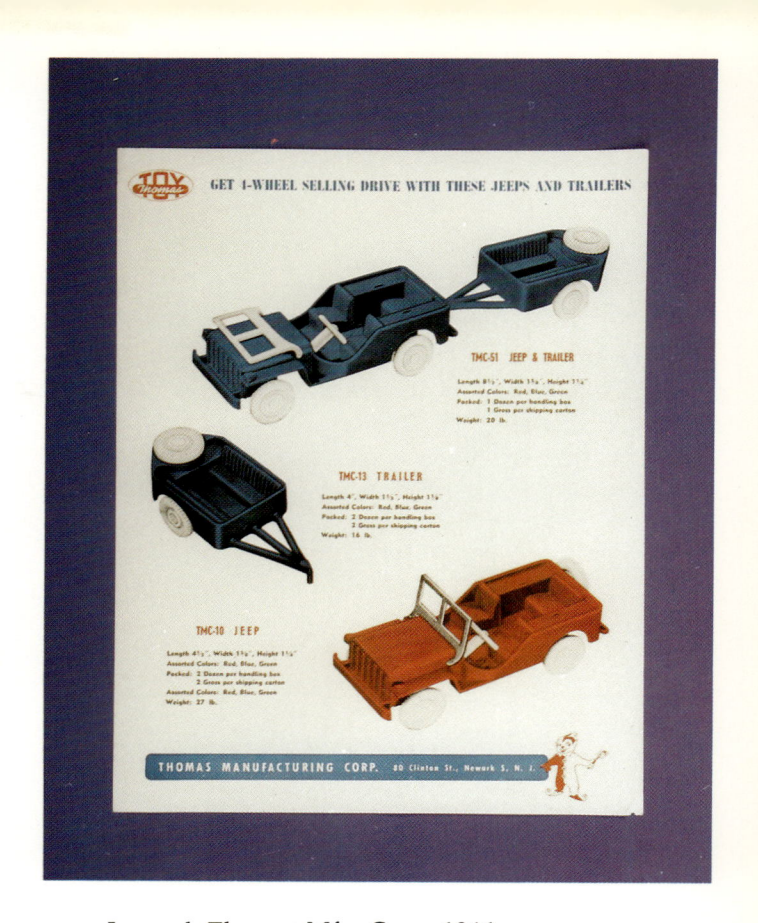

Jeep ad. Thomas Mfg. Corp. 1946.

Jeep, 5-1/2" x 2-1/4" x 2-3/4", with folding windshield and tailgate that drops, assorted color combinations, plastic wheels, California Moulders, Inc., USA, 1947 to early 1950s.

JEEPS 71

Jeep, 5-1/2" x 2-1/4" x 2-3/4", with driver, passenger and folding windshield, assorted color combinations, rubber wheels, California Moulders, Inc., USA (No. CM120), early 1950s.

Jeep Puzzle, 5-1/4" x 2-5/8" x 2-1/2", assorted color combinations, plastic wheels, Character Molding Corp., USA (No. J-69), early to mid 1950s. Suggested Retail $0.69.

Willys Station Wagon, 3-1/2" x 1-1/2" x 1-5/8", assorted color combinations, plastic wheels, Dillon Beck Mfg. Co., USA, early 1950s.

Jeep, 4" x 1-1/2" x 1-1/2", assorted color combinations, plastic wheels, 1948 to early 1950s. In 1948 the mold for the original Ideal jeep was sent to Mexico. It was modified and this is the result.

72 PLASTIC TOYS

Jeep, 5" x 2-3/4" x 1-3/4", a Marx playset vehicle, assorted colors, plastic wheels, Louis Marx and Co., USA, early 1950s.

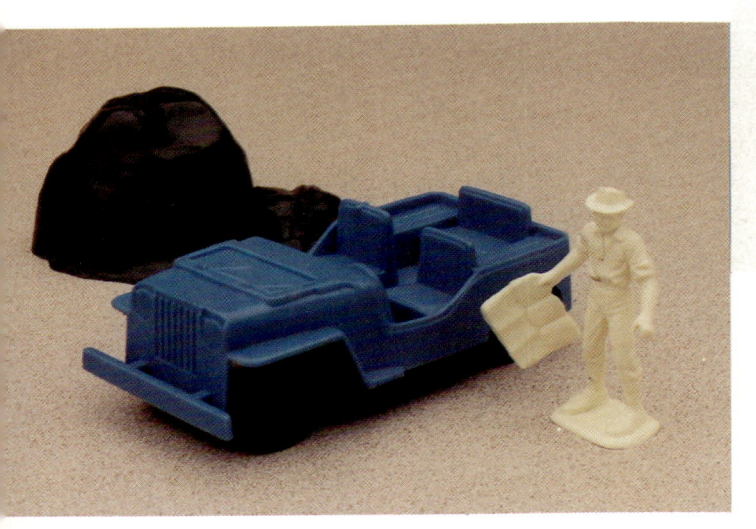

Jeep, 3-3/4" x 1-7/8" x 1-1/4", assorted colors, plastic wheels, Louis Marx and Co., USA, early 1950s.

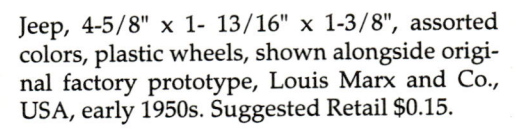

Jeep, 4-5/8" x 1-13/16" x 1-3/8", assorted colors, plastic wheels, shown alongside original factory prototype, Louis Marx and Co., USA, early 1950s. Suggested Retail $0.15.

JEEPS 73

Jeep, 5" x 2-1/2" x 2-1/4", with folding windshield and wind-up motor with attached key, assorted color combinations, rubber wheels, Louis Marx and Co., USA, late 1940s to early 1950s.

Fix-All Jeep, 7" x 4" x 3-3/4", take apart Jeep with over thirty pieces, opening hood and tailgate, folding windshield and removable top, includes the following tools and equipment; spare wheel and tire, gas can, fire extinguisher, adjustable jack, hammer, open end wrench, box wrench, screwdriver and adjustable wrench, assorted color combinations, plastic wheels, Louis Marx and Co., Inc., USA, 1953 to mid 1950s. Suggested Retail $1.79.

Jeep, 5-3/4" x 2-5/8" x 2-1/2", with folding windshield, assorted color combinations, plastic wheels, The Plas-Tex Corp., USA, 1948 to early 1950s.

Jeep and Trailer, 7-3/4" x 1-5/8" x 1-5/8", with attached trailer, assorted color combinations, plastic or rubber wheels, Thomas Mfg. Corp., USA (No. 19), 1946 to mid 1950s, driver added in 1950. Suggested Retail $0.29. Jeep alone (No. 17), 1945 to mid 1950s, driver added in 1950.

Jeep, 3" x 1-3/8" x 1-1/8", assorted colors, plastic wheels, Manufacturer unknown, late 1940s to early 1950s.

JEEPS 75

CHAPTER 8
RACE CARS, HOT RODS AND MOTORCYCLES

With the development of dramatically more powerful automobiles, the question of which manufacturer made the fastest car and who was the best driver were often the subject of hot debate in post-war America. Americans with more leisure time on their hands than any previous generation flocked to racetracks across the country by the millions in the late forties and fifties, to watch their favorite cars and drivers compete in everything from demolition derbies to the Indianapolis 500.

Hot rods and motorcycles were equally as popular and were promoted by groups like the National Hot Rod Association and eulogized in movies like, *Rebel Without a Cause* and *The Wild One*.

Since the toy world reflects the grown-up world so faithfully, it was only natural that race cars, hot rods and motorcycles would become popular playthings.

Of all the companies that produced the aforementioned toys in plastic, Nosco stands out above the rest.

Nosco actually started out as the National Organ Supply Co. of Erie, Pennsylvania, founded in 1920 by Harry H. Kugel. The company manufactured metal organ pipes, parts and supplies for commercial pipe organs which were quite popular at the time. With many of the larger pipe organs of the day having several thousand or more pipes, it's easy to see how a business like this could be successful, and successful it was.

Harry Kugel, however, was a progressive thinker and warned of the day when an electric organ would replace the mighty pipe organ. In 1934 his prediction came true. Laurens Hammond, a U.S. inventor, was granted a patent for the first commercially practical electric organ.

The following year, looking for diversification, the National Organ Supply Co. entered the plastic injection molding field as a custom molder.

The new venture would be called the Nosco Plastics Division of the National Organ Supply Company. The name Nosco was coined because it was felt plastic buyers would have a hard time tying plastics manufacturing to the National Organ Supply Company.

Harry's brother, Reuben G. Kugel, joined the company in the early 1940s as Treasurer and later served as Executive Vice President when Harry passed away in 1957.

Nosco's toy division appears to have started after the war according to the late Reuben Kugel's son, Charles.

In 1948 Nosco introduced an innovative four inch long wind-up bus and "woodie" station wagon that used plastic gears and keys instead of ones made of metal. The success of these two vehicles must have convinced Nosco to expand its toy line for 1949.

For 1949, a smaller series of plastic wind-up vehicles about three and one-half inches long and incorporating

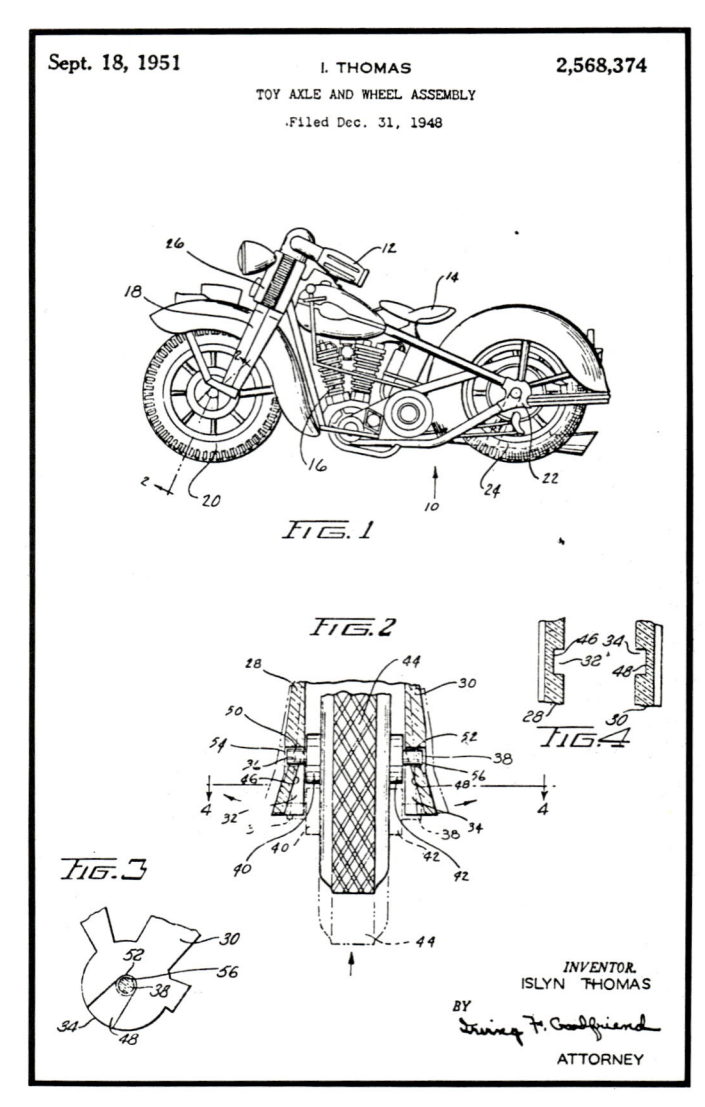

Motorcycle patent drawing. Thomas Mfg. Corp. 1951.

plastic gears and keys was introduced. Called "Auto-Mites", the series included a sedan, coupe, wrecker and a downsized "woodie" station wagon.

The big news of 1949, however, was the introduction of Nosco's nine and one-half inch wind-up "Doodle Bug Midget Racer" and their six inch wind-up "Cop-Cycle." These two toys are considered by most collectors to be the finest examples of a race car and motorcycle produced in plastic.

Nosco would continue to delight its fans by offering their nine and one-half inch friction powered "Hot'See" hopped-up hot rod in 1951 and their nine inch friction powered "Vizy Vee" stock car in 1952.

Both of these "works of art" featured visible working motors which must have seemed very "cool" at the time. Both are now considered to be the ultimate plastic hot rods.

While Nosco was releasing what seemed like one hit after another, the majority of its toy business was molding plastic novelties for Cracker Jack, according to Charles Kugel. This previously unknown side of Nosco helps explain the company's tremendous success in the late forties and early fifties.

Success is often fleeting though and Nosco's toy division would eventually be sold to Saunders Tool and Die Co. of Aurora, Illinois in 1955, never to be heard from again.

Race Car, 3-1/2" x 1-3/4" x 7/8", assorted colors, plastic wheels, B. W. Molded Plastics, USA, early 1950s. Suggested Retail $0.10.

Race Cars, 2-1/2" x 3/4" x 3/4", assorted colors, plastic wheels, Dillon Beck Mfg. Co., USA, 1948 to early 1950s. Suggested Retail $0.05.

Speed-Kop Whistle, 4" x 1" x 2-5/8", makes siren sound when blown and wheels appear to turn, assorted colors, Commonwealth Plastics Corp., USA (No. T117), 1949 to early 1950s. Suggested Retail $0.10.

Jet Racer, 5-1/4" x 2-1/4" x 1-3/4", spring powered catapult action, assorted colors, plastic wheels, Elmar Products Co., USA, early 1950s.

RACE CARS, HOT RODS AND MOTORCYCLES 77

Stock Car Races, folded card, 6-3/4" x 6-1/2", cars, 2" x 1" x 3/4", assorted colors, Empire Plastic Corp., USA (No. 51), early 1950s.

Race Car, 6" x 3-3/8" x 2-1/4", with driver, assorted color combinations, rubber wheels, Glen Dimension, USA, early 1950s.

Stock Car Races, expanded card, 24" x 6-1/2", Empire Plastic Corp., USA (No. 51), early 1950s.

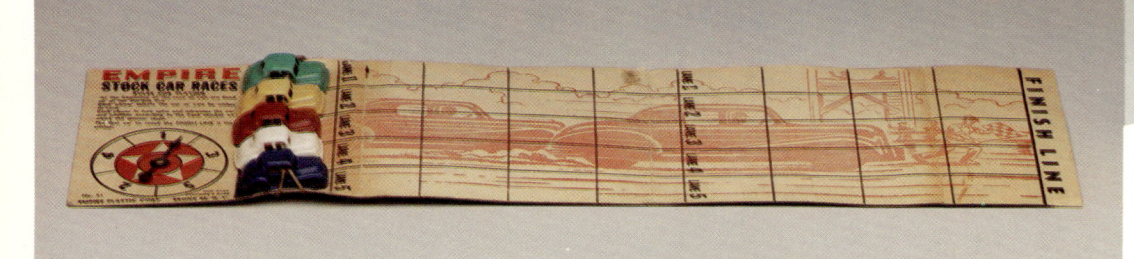

Motorcycle Puzzle, assorted color combinations, Manufacturer Unknown, USA, early 1950s. Motorcycle Whistle, 3-1/2" x 7/8" x 2", when whistle is blown a siren sounds, assorted colors, Gerber Plastic Co., USA, 1949 to early 1950s. Suggested Retail $0.10.

Motorcycle, 5" x 1-3/4" x 3-1/4", with removable cop, assorted colors, rubber wheels, Hubley Mfg. Co., USA, 1949 to mid 1950s.

Motor Scooter, 4" x 2" x 2-7/8", handlebars turn front wheel and luggage compartment opens, assorted color combinations, plastic or rubber wheels, Ideal Novelty and Toy Co., USA (No. MS-50), 1948-1950. Suggested Retail $0.49.

Racer, 4-1/2" x 1-5/8" x 1-3/8", assorted colors including vacuum metalized, plastic wheels, Ideal Novelty and Toy Co., USA (No. RC-10), 1948-1949. Suggested Retail $0.10.

Mechanical Racing Car, 8-1/4" x 4" x 2-3/4", wind-up motor, assorted color combinations, white rubber wheels, Ideal Novelty and Toy Co., USA (No. RCM-120), 1948-1950. Suggested Retail $1.19.

RACE CARS, HOT RODS AND MOTORCYCLES 79

Mechanical Motorcycle with Sidecar and Cop, 3-11/16" x 5-9/16" x 6-5/16", wind-up with attached key, detachable cop, assorted color combinations, rubber tires on plastic hubs, Ideal Novelty and Toy Co., USA (No. MC-120), 1950. (Left hand is not broken off but molded that way).

Mechanical Motorcycle with Sidecar and Cop, detail, Ideal Novelty and Toy Co., USA (No. MC-120), 1950.

Dream Racer, 13-1/2" x 5-1/2" x 2-3/4", with detailed visible chrome finish motor, black with gold paint, plastic wheels, Ideal Toy Corporation, USA, (No. 3043), 1954-1955. Suggested Retail $1.50.

Race Car, 6-1/4" x 2-1/4" x 1-5/8", with No. 4 on hood, with or without driver, assorted colors, rubber wheels, Lido Toy Corp., USA, 1949 to mid 1950s. Suggested Retail $0.10.

"Speedy Pete" Racer Pull Toy, 16" long; when pulled along, wheels produce a clicking sound and driver's head bobs up and down on a spring; lithographed cardboard cylinder with assorted color plastic ends, driver's head and wheels, Ideal Toy Corporation, USA, (No. 4501), 1955. Suggested Retail $1.20.

80 PLASTIC TOYS

Stock Car, 11" x 4" x 3-1/2", with opening hood and trunk with spare tire, jack and lug wrench; makes motor sound when pushed; red, yellow and black with assorted dents and cut out fender wheels, decals and No. 77 painted on side, rubber tires on metal wheels, Lincoln Line, Inc., USA, 1952 to mid 1950s. Suggested Retail $1.80.

Stock Car Racer, 5" x 2-3/8" x 1-7/8", with back up recoil motor and No. 8 on back, assorted colors, rubber wheels, Lincoln Line, Inc., USA, early 1950s. Suggested Retail $0.25.

Stock Car Racer, 3-1/8" x 1-1/4" x 1-1/4", a Marx playset vehicle, assorted colors, metal wheels, Louis Marx and Co., USA, early to mid 1950s.

Take-A-Part Hot Rod, 3-1/2" x 1-1/2" x 1-1/8", assorted colors, plastic wheels, Louis Marx and Co., USA, early to mid 1950s.

RACE CARS, HOT RODS AND MOTORCYCLES

"Cop-Cycle," 6" x 3-1/2" x 4", friction motor, assorted color combinations, rubber rear wheels, plastic front wheel, Nosco Plastics, USA (No. 6357), 1949 to early 1950s.

"Cop-Cycle," detail, Nosco Plastics, USA (No. 6357), 1949 to early 1950s.

"Doodle Bug Midget Racer," 9-1/2" x 5-3/8" x 4", with wind-up motor and attached key, assorted color combinations with number decals, rubber tires on plastic hubs, Nosco Plastics, USA (No. 6381), 1949 to early 1950s. Suggested Retail $1.50.

Ace Indianopolis Mechanical Racer, 8" x 4-1/4" x 3", with wind-up motor and attached key, assorted colors with silver painted details and No. 5 on hood, Nosco Plastics, USA (No. 6390), 1949 to mid 1950s. Suggested Retail $0.98.

82 PLASTIC TOYS

"Hot'See Hopped-up Hot Rod", 9-1/2" x 4-3/4" x 4", with friction motor and driver, visible in-line four-cylinder motor with working pistons, fan and removable body, assorted color combinations, rubber tires on plastic wheels, Nosco Plastics, USA (No. 6490), 1951 through early 1950s. Suggested Retail $2.98.

"Hot'See Hopped-up Hot Rod", detail, Nosco Plastics, USA (No. 6490), 1951 through early 1950s.

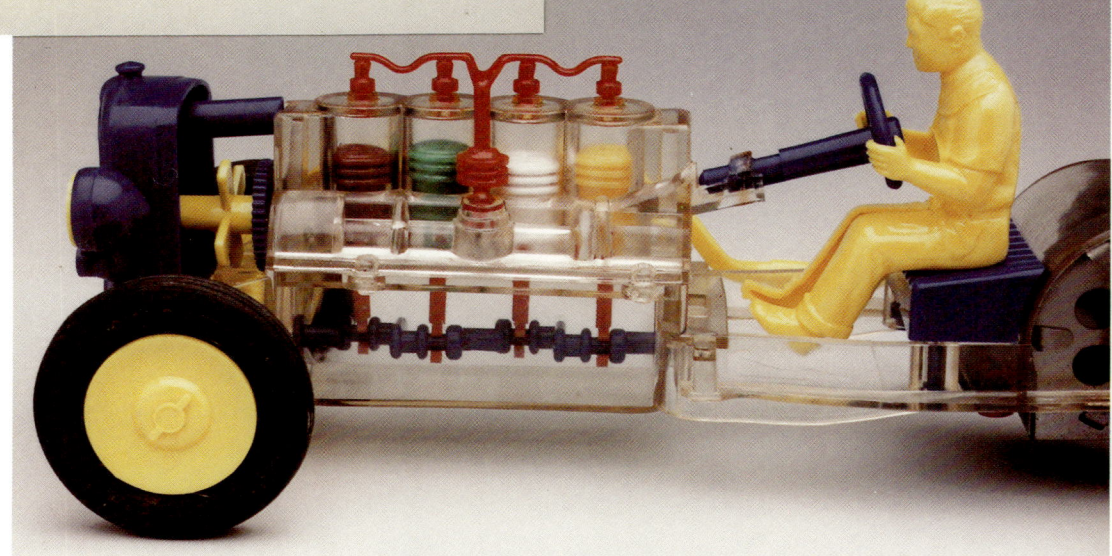

"Hot'See Hopped-up Hot Rod", box 10-1/4" x 5" x 4-1/4", Nosco Plastics, USA (No. 6490), 1951 through early 1950s.

"Vizy Vee Stock Car Racer", 9" x 3-1/4" x 3-1/2", with friction motor and visible V8 with working pistons, assorted color combinations, rubber tires on plastic wheels, Nosco Plastics, USA (No. 6564), 1952 through early 1950s. Suggested Retail $2.98.

RACE CARS, HOT RODS AND MOTORCYCLES 83

Motor Bike Racer, 5" x 2-1/2" x 4-1/4", with wind-up motor, attached key and rider, white with red painted details, plastic wheels, Precision Plastics Co., USA (No. MB-80), 1948 to mid 1950s.

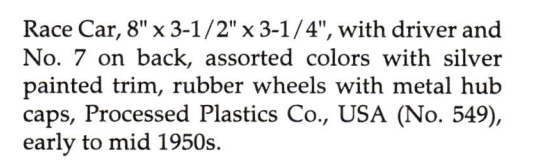

Race Car, 8" x 3-1/2" x 3-1/4", with driver and No. 7 on back, assorted colors with silver painted trim, rubber wheels with metal hub caps, Processed Plastics Co., USA (No. 549), early to mid 1950s.

Hot Rod, 5" x 2-1/4" x 2", with driver, assorted colors, rubber wheels, Processed Plastics Co., USA, early 1950s.

Motorcycle with Sidecar, 3-3/8" x 1-1/4" x 1-1/2", with attached baby in sidecar, assorted color combinations, plastic wheels, Pyro Plastics Corp., USA, early to mid 1950s. Suggested Retail $0.25.

84 PLASTIC TOYS

Race Car, 3-3/4" x 1-3/4" x 1-3/4", with driver and No. 7 on hood, assorted colors, plastic wheels, Pyro Plastics Corp., USA, early 1950s. Suggested Retail $0.10.

Race Car, 3-3/4" x 1-3/4" x 1-3/4", with driver and No. 2 on hood, assorted colors, plastic wheels, Pyro Plastics Corp., USA, early 1950s. Suggested Retail $0.10.

Supersonic Soap Box Racer, 4-3/4" x 2-1/2" x 2-3/4", with driver and No. 7 on back, assorted colors, plastic wheels, Pyro Plastics Corp., USA, early 1950s.

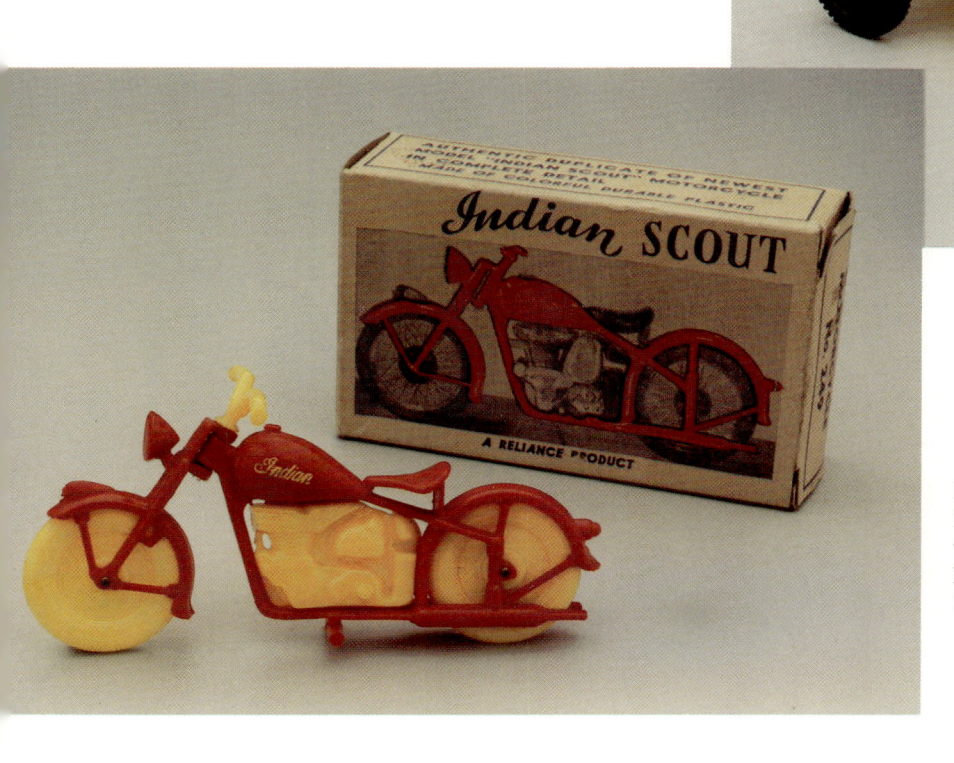

Indian Scout Motorcycle, 5" x 2" x 2-1/2", handlebars turn front wheel, red and yellow with yellow hot stamping, plastic wheels, Reliance Molded Plastics, Inc., USA (No. 249), 1949 to early 1950s.

RACE CARS, HOT RODS AND MOTORCYCLES 85

Racer, 4-3/4" x 1-3/4" x 1-1/4", assorted colors, plastic wheels, Renwal Manufacturing Co., Inc. USA (No. 61), with painted chrome trim (No. 2061), 1948 without driver, 1949-1956 with driver. Suggested Retail $0.10.

Racer, 4-3/4" x 1-3/4" x 1-1/4", assorted colors, plastic wheels, Renwal Manufacturing Co., Inc. USA (No. 88), with painted chrome trim (No. 2088), 1949-1954. Suggested Retail $0.10.

Motorcycle with Sidecar, detail, Renwal Mfg. Co., Inc., USA (No. 23), 1949-1954.

Motorcycle with Sidecar, 5-1/4" x 3-3/8" x 2-5/8", with passenger only, handlebars steer front wheel, assorted color combinations, plastic wheels, Renwal Manufacturing Co., Inc. USA (No. 23), 1949-1954. Suggested Retail $0.39.

86　PLASTIC TOYS

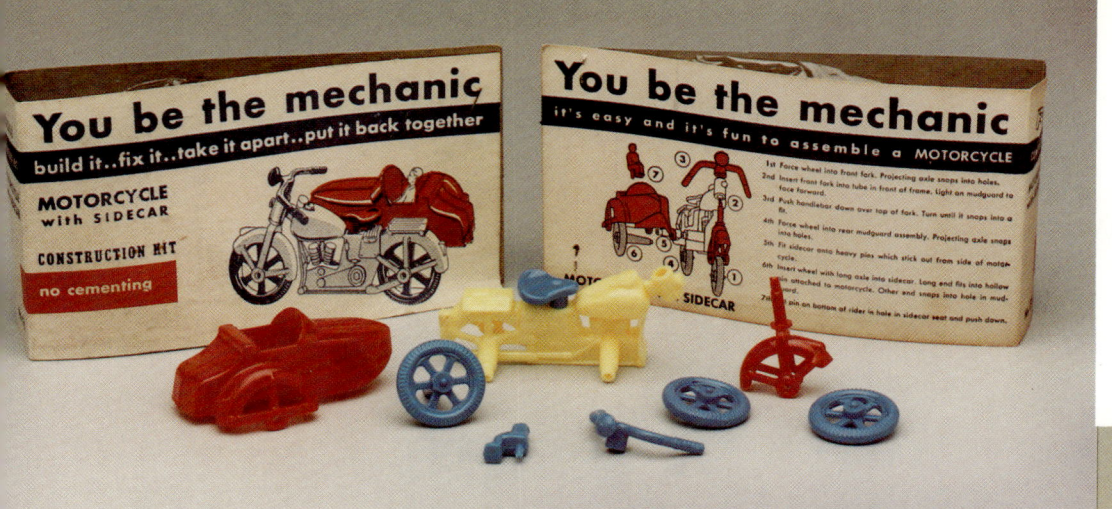

Motorcycle with Sidecar Construction Kit, Renwal Mfg. Co., Inc., USA (No. 175), 1953-1955.

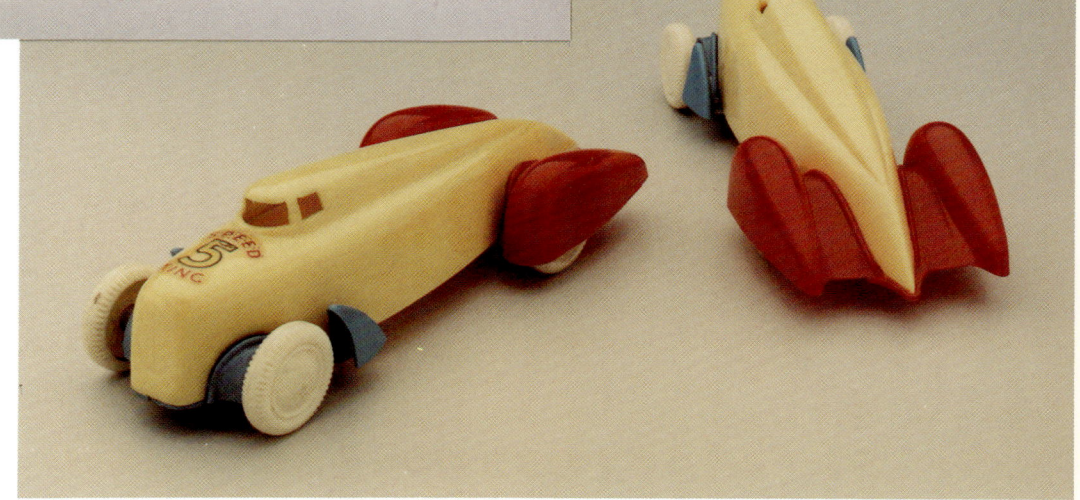

Racing Car, 6-3/8" x 2-3/8" x 2", assorted color combinations with a number decal on hood, plastic wheels, Renwal Manufacturing Co., Inc. USA (No. 58), 1949-1954. Suggested Retail $0.29.

Speed King Racer, 10-1/2" x 3-3/4" x 3-1/8", with driver, friction motor, assorted color combinations, rubber tires on plastic hubs, Renwal Manufacturing Co., Inc. USA (No. 107), without friction motor and with plastic wheels (No. 207), 1950-1951 and 1952-1954 without motor. Suggested Retail $0.79.

Racer, 3-1/4" x 1-1/8" x 3/4", assorted colors, plastic wheels, Renwal Manufacturing Co., Inc. USA (No. 150), 1950-1955. Suggested Retail $0.05.

RACE CARS, HOT RODS AND MOTORCYCLES 87

Motorcycle Cop, 3-3/4" x 2-1/4" x 1", motorcycle and cop stand upright and roll on floor without falling over, removable cop, assorted color combinations, plastic wheels, Renwal Manufacturing Co., Inc. USA (No. 189), 1953-1955. Suggested Retail $0.15.

Motorcycle Cop, 9" x 5-1/2" x 2-1/4", motorcycle and cop stand upright and roll on floor without falling over, removable cop, assorted color combinations, plastic wheels, Renwal Manufacturing Co., Inc. USA (No. 188), 1953-1955. Suggested Retail $0.69.

World Champion Racer, 10-3/8" x 4" x 3-3/8", with friction motor and driver, assorted colors, rubber wheels, Renwal Mfg. Co., Inc., USA (No. 216), 1954-1955. Suggested Retail $1.49.

Race Cars, red with driver, 3" x 1" x 3/4", silver with driver 2-3/8" x 1-3/8" x 1-1/8", blue with driver 2-3/4" x 1-1/4" x 3/4", assorted colors, plastic wheels, Ross Tool and Mfg. Co., USA, 1949 to early 1950s.

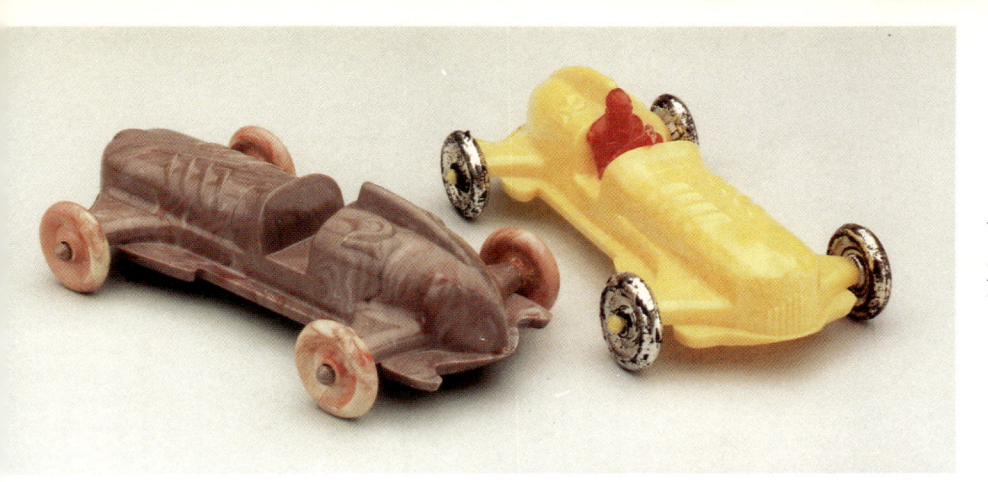

Race Car, 4" x 2-1/8" x 1-1/4", with No. 2 on back, assorted colors, plastic wheels, Ross Tool and Mfg. Co., USA (design patent No. D.135458), 1949 to early 1950s.

Hot Rod, 7" x 3-5/8" x 3", with friction motor and driver, red with painted details and number decal, rubber tires on metal wheels, Saunders Tool and Die Co., USA (No. 400), 1950 to mid 1950s. Suggested Retail $0.98.

Motorcycle and Rider, 4" x 1-1/2" x 3", with painted vinyl detachable rider and handlebars that steer front wheel, assorted color combinations, plastic wheels, Thomas Manufacturing Corp., USA (No. 72), 1949-1951, (No. 107) 1952 to mid 1950s, also available in olive drab with detachable military policeman, (No. 125) if vacuum metalized, sold with and without plated rider, 1950 to mid 1950s.

Motorcycle and Rider, detail vacuum metalized version, Thomas Mfg. Corp., USA (No. 125), 1950 to mid 1950s.

RACE CARS, HOT RODS AND MOTORCYCLES

Service Motorcycle and Rider, 4-7/8" x 2-1/8" x 2-3/4", with painted vinyl detachable rider, rear compartment that opens and handlebars that steer front wheel, assorted colors, plastic wheels, Thomas Manufacturing Corp., USA (No. 90), 1950 to mid 1950s.

Midget Racer, 1-5/8" x 3/4" x 5/8", assorted color combinations, stationary plastic wheels, Thomas Mfg. Corp., USA (No. 141), 1951-1952. Suggested Retail $0.05. Also sold with truck (No. 41) as Truck and Racer (No. 132), 1951-1952. Suggested Retail $0.15.

International Racer, 5" x 2-1/4" x 1-3/4", with No. 5 on hood, assorted color combinations, plastic wheels, Thomas Toys, USA, 1955 to late 1950s. Suggested Retail $0.10.

90 PLASTIC TOYS

Motorcycle, 4" x 1-1/2" x 3", with removable rider and wind-up motor, with removable crank to wind, red and white, plastic wheels, (possible mail-in cereal premium), Manufacturer Unknown, USA, early 1950s.

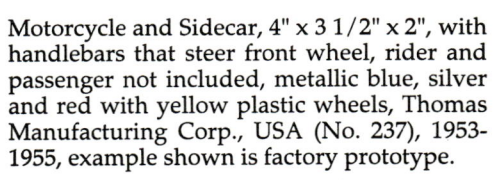

Motorcycle and Sidecar, 4" x 3 1/2" x 2", with handlebars that steer front wheel, rider and passenger not included, metallic blue, silver and red with yellow plastic wheels, Thomas Manufacturing Corp., USA (No. 237), 1953-1955, example shown is factory prototype.

Motorcycle and Sidecar Detail, 4" x 3 1/2" x 2", Thomas Manufacturing Corp., USA (No. 237), 1953-1955.

Motorcycle, 4 1/2" x 1 1/4" x 2 1/8", with attached rider and wind-up motor with attached key, red, plastic wheels, European, early 1950s.

Balloon Jet Racer, 4-3/4" x 1-1/2" x 1-1/2", balloon powered, assorted colors, plastic wheels, Manufacturer unknown, USA, early 1950s.

RACE CARS, HOT RODS AND MOTORCYCLES 91

CHAPTER 9
TRUCKS, CONSTRUCTION VEHICLES AND BUSES

The post-war years were a boom for the construction trade with 1.396 million new homes started in 1950 alone. Everywhere children looked, they saw dump trucks, steam shovels and bulldozers building roads, and clearing and leveling sights for new construction. The suburbs had arrived!

New homes and businesses required new services and as soon as the construction equipment rolled out, the fire trucks, delivery vans, ambulances, buses and tow trucks rolled in.

The children were fascinated by these shiny, powerful mobile giants and dreamed of the day when they could race to a fire aboard a half block long hook and ladder truck or operate their own fleet of heavy equipment.

Toy manufacturers have always tried to satisfy the fascination of youth and were quick to capitalize on an insatiable appetite for trucks, trucks and more trucks.

Plastic Company of Aurora, Illinois, one of the few companies represented in this book that is still in business today. Its founder, Ross Bergman, started out as a teacher and later became a high school principal in Aurora during the Second World War. After the war, Bergman began to consider alternative forms of employment that might better support his growing family.

In 1946, while playing golf, he met Paul Saunders, owner of Saunders Tool and Die Co., also of Aurora. Saunders had a successful metal stamping business during the war and now, like many other tool and die shops, wanted to try his hand at manufacturing plastic toys that would incorporate his metal fabricating skills. With his years of working with and teaching children, Bergman seemed qualified for selling toys and was hired as Sales Manager of Saunder's new venture.

Saunder's first toys were a wind-up Indy-type racer, a wind-up sedan and a toy bugle. The early success of

Express Truck, 4" x 1-3/8" x 1-3/8", Kilgore Mfg. Co., USA, 1938 to late 1940s.

The only pre-war injection molded plastic toy truck was manufactured by the Kilgore Mfg. Co. of Westerville, Ohio. It was a four inch long express truck that was part of Kilgore's "Jewels For Playthings" line first offered in 1938. The first plastic toy bus was also part of the Kilgore line, along with a sedan, coupe, taxi and airplane.

At present, there is no evidence of any of the "Jewels For Playthings" being produced during the war. Part of the line may have been produced for a very short time after the war, makin these vehicles highly sought after by today's plastic collectors.

Ideal was the next company to produce an injection molded truck and added a pickup to their line in 1945. Banner, Dillon Beck and Thomas added trucks to their lines in 1946 while Renwal and Processed Plastic introduced them in 1948.

It was, in fact, a toy truck that started the Processed

Saunder's toy division was a credit to the hard work and persuasive personality of Bergman. All was not a bed of roses, however, and after two years and a disagreement with Saunders over Bergman's future and the future of the company, Bergman left.

Not one to be discouraged, Bergman knew he had found his niche and used the money he had saved to have the mold for a toy dump truck made. With this as his only toy, Bergman found two partners and founded the Processed Plastic Co. on March 1, 1948. The little eight inch dump truck was an immediate hit, with Woolworth's and Kresge's placing large orders.

The rest of the Processed Plastic Company's success story is typical of American free enterprise and individual initiative. Bergman eventually bought out his partners, and his two sons, Robert and David, joined the business in 1961. Today, under the guidance of Robert

Dump Truck, 8-1/2" x 2-3/4" x 2-3/4", Processed Plastics Co., USA, 1948 to early 1950s.

and David, the company offers over one hundred and fifty items ranging from riding toys to miniature figures.

Manufacturers like the Processed Plastic Co. loved toy trucks as much as their customers did, because once they had the mold for a basic chassis and cab all they had to do was mold new bodies which saved them time and thousands of dollars. Dillon Beck was among the first to take advantage of this with their best-selling ten cent dump truck, cement mixer and steam shovel, all using the same chassis and cab.

In 1946, Banner and Dillon Beck introduced the first plastic toy dump trucks and fire trucks. Banner also introduced the first trailer truck, wrecker and oil truck. Ideal added a nifty tear-drop shaped utility trailer for the pickup and sedan they had introduced in 1945.

In 1947, Ideal introduced the first plastic toy cement mixer truck and farm tractor. Ideal and Thomas added oil trucks to their lines and Thomas added a wrecker, complete with operating hoist and crank.

In 1948, Banner Plastics introduced the first plastic toy steam roller, and California Moulders, the first plastic toy steam shovel. Ideal introduced the first plastic toy steam shovel truck and the first sanitation truck. Renwal introduced the first plastic toy truck cab with opening doors.

In 1949, Renwal introduced the first plastic toy auto carrier, complete with four autos. Ideal introduced the first panel truck with rear doors that opened and closed.

In 1950, Renwal introduced the first true plastic toy hook and ladder truck that actually steered from the rear like a real one! Ideal introduced the first plastic toy ambulance and bulldozer.

From 1950 on, there seemed to be two schools of thought concerning plastic toy trucks. The first school said keep 'em small and on the ten to twenty-nine cent counter. This group included manufacturers like Allied, Banner, Dillon Beck, Gilmark, Pyro and Thomas. The second school of thought said make 'em every size imaginable and this group included manufacturers like Ideal, Hubley, Marx, Processed Plastic and Renwal.

As the 1950s drew to a close, foreign competition began to take its toll on all but the manufacturers from the second group, who had invested in larger, more expensive toys. They managed to keep the wolves from the door awhile longer but, in the end, all but the Processed Plastic Co. knuckled under.

Dump Truck, 3-1/2" x 1-1/4" x 1-1/4", 1946 to mid 1950s, Cement Mixer, 3-1/4" x 1-5/16" x 1-3/4", 1948 to mid 1950s, Steam Shovel, 4" x 1-5/16" x 1-3/4", 1949 to mid 1950s, Dillon Beck Mfg. Co., USA.

TRUCKS, CONTRUCTION VEHICLES AND BUSES

Cement Mixer, 4" x 2-5/8" x 3", turn crank to rotate drum, assorted color combinations, plastic wheels, All Metal Products Co., USA, 1949 to early 1950s; Wheelbarrow, 3-5/8" long, Commonwealth Plastics Corp., USA, 1949 to early 1950s.

Ladder Fire Truck, 5-3/4" x 1-5/8" x 1-3/8", ladder swivels and raises, red, plastic wheels, All Metal Products Co., USA, early 1950s.

Steam Roller, 3-3/4" x 1-1/2" x 1-5/8", assorted color combinations, plastic wheels and roller, Allied Molding Corp., USA, early 1950s.

Cement Mixer, 2-1/4" x 1-1/4" x 1-1/4", assorted color combinations, plastic wheels, Allied Molding Corp., USA, early 1950s.

94 PLASTIC TOYS

Moving and Furniture Van, 5-1/2" x 1-5/16" x 1-7/8", with miniature furniture, Allied Molding Corp., USA, early 1950s. Suggested Retail $0.59.

Sand Set, box 12" x 2-3/4" x 2", includes 3" long trailer with 2-1/4" long car, 3-1/2" long cement mixer, and 4" long dump truck, assorted color combinations, plastic wheels, Allied Molding Corp., USA (No. 600), early 1950s.

Steam Shovel, 6-3/4" x 1-7/8" x 3", with working scoop and cab that swivels, assorted color combinations, plastic wheels, Allied Molding Corp., USA, early 1950s.

Auto Carrier, 9-1/2" x 1-11/16" x 1-11/16", with four 2-1/4" long cars with stationary wheels and an unloading ramp that folds down, assorted color combinations, plastic wheels, Allied Molding Corp., USA, early 1950s.

TRUCKS, CONTRUCTION VEHICLES AND BUSES 95

Fire Chief's Convertible, 5" x 1-7/8" x 1-1/2", with driver, Ladder Fire Truck, 6-1/2" x 1-3/4" x 1-3/4", with driver, 4 firemen and removable ladder, vacuum metalized, plastic wheels, Allied Molding Corp., USA, early 1950s.

Steam Roller, 4-1/2" x 2-3/4" x 3", rubberband motor, red, yellow, and blue, plastic wheels and roller, Archer Plastics, USA, early 1950s.

Fire Truck, 3-1/2" x 1-1/4" x 1-3/16", assorted colors, plastic wheels, B. W. Molded Plastics, USA, early 1950s.

Pickup Truck, 3-1/2" x 1-1/4" x 1-3/8", assorted colors, plastic wheels, B. W. Molded Plastics, USA, early 1950s.

96 PLASTIC TOYS

Nozzle Truck, 5-3/8" x 1-3/4", Ladder Truck, 5-3/8" x 1-3/4" x 1-3/4", red and white, plastic wheels, Bachmann Inc., USA, early 1950s to 1970's.

Bus, 5" x 1-1/2" x 1-3/4", assorted colors, plastic wheels, Bachmann Brothers Inc., USA, 1954 to 1960's.

Ladder Fire Truck, 4-3/8" x 1-3/4" x 1-1/4", plastic or tin chassis, assorted colors, plastic wheels, Banner Plastics Corp., USA (No. 103), 1948 to early 1950s. Suggested Retail $0.10.

Dump Truck, 4-5/8" x 1-1/4" x 1-3/8" (No. 102), Farm Truck, 4-1/4" x 1-1/2" x 1-1/2" (No. 104), Pickup Truck, 4-3/8" x 1-5/8" x 1-1/4" (No. 105), Oil Truck, 4-3/8" x 1-3/4" x 1-1/2" (No. 101), assorted color combinations, plastic wheels, Banner Plastics Corp., USA, 1948 to early 1950s. Suggested Retail $0.10.

TRUCKS, CONSTRUCTION VEHICLES AND BUSES

Sand Loader, 3" x 1-1/2" x 3-1/4", with working conveyor and rubber wheels on plastic treads; Steam Shovel, 4-1/2" x 1-5/8" x 2-5/8", with working scoop and plastic treads, assorted color combinations, Banner Plastics Corp., USA, 1948 to early 1950s. Suggested Retail $0.25.

Trailer Truck, 7-1/4" x 1-3/4" x 2-5/8", assorted color combinations, plastic wheels, Banner Plastics Corp., USA (No. 69), 1946. Suggested Retail $0.29.

Delivery Van, 4-3/8" x 1-5/8" x 2", assorted color combinations, plastic wheels, Banner Plastics Corp., USA, 1949 to early 1950s. Suggested Retail $0.10.

Nozzle Truck, 5-3/8" x 1-13/16" x 2-1/4", with wind-up motor and attached key, assorted color combinations, plastic wheels, Banner Plastics Corp., USA, 1949. Suggested Retail $0.29.

Earth Hauler, 4-1/2" x 1-5/8" x 1", Road Grader, 4-1/2" x 1-1/2" x 1-1/4", assorted colors, plastic wheels, Banner Plastics Corp., USA, 1949 to early 1950s.

Equipment Hauler, tractor trailer 9-1/4" x 1-1/2" x 2", boat, 4" x 1-1/4" x 1-3/16", assorted color combinations, plastic or rubber wheels, Banner Plastics Corp., USA, 1949 to mid 1950s. Suggested Retail $0.29. Other loads included a tractor and a steam roller. Boat alone 1946 to mid 1950s. Suggested Retail $0.10.

Trailer Truck, 7" x 1-3/4" x 2-1/2", Gasoline Truck, 7" x 1-3/4" x 1-3/4", assorted colors and color combinations, plastic wheels, Banner Plastics Corp., USA, 1949 to mid 1950s. Suggested Retail $0.29.

Cross Country Express Livestock Truck, 5-1/4" x 1" x 1-3/8", Cross Country Express Gasoline Truck, 5-1/4" x 1" x 1-3/8", assorted colors and color combinations, plastic wheels, Banner Plastics Corp., USA, 1950 to mid 1950s, additional trailers included coal and enclosed.

TRUCKS, CONTRUCTION VEHICLES AND BUSES 99

Cement Mixer, 4" x 1-3/8" x 1-5/8", Stake Bed Truck, 4-5/8" x 1-3/4" x 1-1/2", Dump Truck, 4-3/4" x 1-1/2" x 1-1/4", Sanitation Truck, 4-1/2" x 1-3/8" x 1-1/2", assorted colors, plastic wheels, Banner Plastics Corp., USA, early to mid 1950s. Suggested Retail $0.10.

Cement Mixer, 4" x 2" x 2-1/2", turn crank to rotate drum, assorted color combinations, plastic wheels, Banner Plastics Corp., USA, early 1950s.

Dump Truck, 6" x 2-1/8" x 2-1/2", with removable snow plough; when lever is pulled, bed lifts and tailgate swings open; Sanitation Truck, 6-1/2" x 2-1/4" x 3", with removable snow ploug;, when lever is pulled, dumpster lifts and empties; assorted color combinations, rubber wheels, Banner Plastics Corp., USA. 1952 to mid 1950s. Suggested Retail $0.29. Also sold without snow plough.

Ladder Fire Truck, 6-1/4" x 2" x 2", with separate windshield, headlights and removable ladder, red plastic wheels, 1947-1948. Pumper Fire Truck, 6-1/4" x 2" x 2", removable ladder, assorted colors, rubber wheels, California Moulders, Inc., USA (No. CM 130), 1949 to early 1950s.

Aerial Ladder Fire Truck, 8" x 2" x 2-1/2", with 2 firemen, 2 extension ladders and main ladder that swivels and raises, red, rubber wheels, California Moulders, Inc., USA (No. 250), 1949 to early 1950s. Suggested Retail $0.39. Also sold with separate headlights, windshield, plastic wheels and without firemen, 1947-1948.

Stake Bed Truck, 5-3/4" x 1-7/8" x 1-3/4", with removable stakes, separate windshield, headlights and taillights, red, plastic wheels, California Moulders, Inc., USA (No. CM 100), 1947 to 1948, also made without separate windshield, headlights, no taillights and with rubber wheels, 1949 to early 1950s.

Ladder Fire Truck Whistle, 4-1/2" x 3/4" x 1", makes siren sound when blown, assorted colors, Commonwealth Plastics Corp., USA, 1949 to early 1950s. Suggested Retail $0.10.

Dump Truck, 3-1/2" x 1-1/4" x 1-1/4", with bed that dumps, (No. D-2), 1946 to mid 1950s; Steam Shovel, 4" x 1-5/16" x 1-3/4", with rotating cab and working shovel, (No. SS-2), 1949 to mid 1950s; Cement Mixer, 3-1/4" x 1-5/16" x 1-3/4", with revolving drum that lifts to unload, (No. M-2), 1948 to mid 1950s, assorted color combinations, plastic wheels, Dillon Beck Mfg. Co., USA Suggested Retail $0.10 each.

TRUCKS, CONSTRUCTION VEHICLES AND BUSES

Ladder Fire Truck, 4-1/2" x 1-1/4" x 1-1/4", ladder raises, (No. F-2), 1946 to mid 1950s. Suggested Retail $0.10; Ladder Fire Truck, 6" x 1-1/2" x 1-3/8", ladder raises, (No. F-3), 1947 to mid 1950s. Suggested Retail $0.25. Both are red with plastic wheels, Dillon-Beck Manufacturing Co., USA.

Power Scoop, 4" x 1-5/8" x 1-1/2", with driver and shovel attachment that works by means of a lever, assorted color combinations, plastic wheels, Dillon Beck Mfg. Co., USA (No. W-100), 1948 to mid 1950s. Suggested Retail $0.10.

Earth Hauler, 4-3/8" x 1-1/4" x 1-3/8", with driver, assorted color combinations, plastic wheels, Dillon Beck Mfg. Co., USA (No. E-2), 1949 to mid 1950s. Suggested Retail $0.10.

Assorted Truck and Trailers, 5" long, assorted color combinations, plastic wheels, Dillon Beck Mfg. Co., USA, 1950 to mid 1950s.

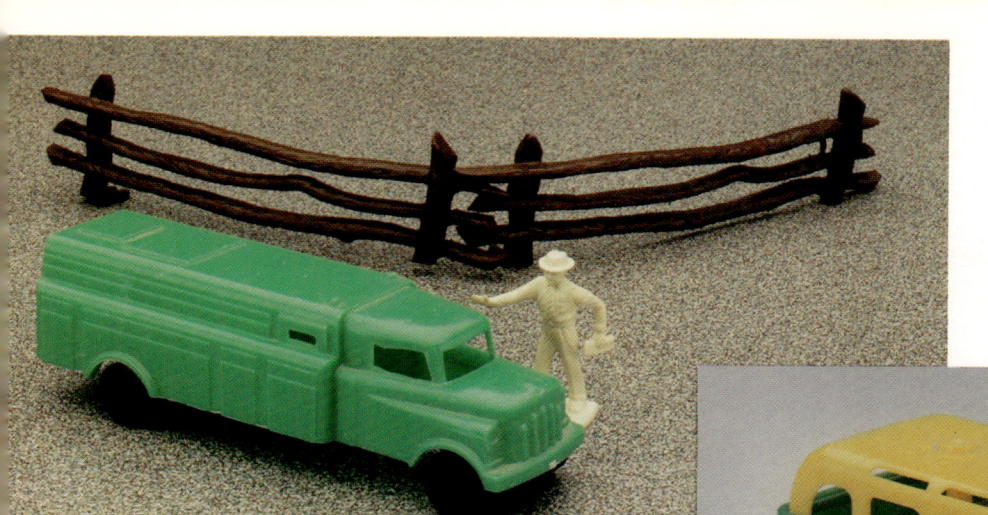

Utility Truck, 5" x 1-9/16" x 1-1/2", assorted colors, plastic wheels, Dillon Beck Mfg. Co., USA, 1953 through mid 1950s. Suggested Retail $0.10.

Bus, 5-3/8" x 1-1/2" x 1-1/2", assorted color combinations, plastic wheels, Dillon Beck Mfg. Co., USA, early to mid 1950s.

Road Builders, card 6" x 4", five piece set includes: steam roller, steam shovel, conveyor loader, bulldozer and dump truck, average length 1-1/2", assorted colors, stationary wheels, Empire Plastics Corp., USA (No. 33), late 1940s to early 1950s.

Assorted small vehicles, 1-1/2" to 2-1/2" long, assorted colors, stationary plastic wheels, Vibro-Roll Products, USA, late 1940s to mid 1950s.

TRUCKS, CONTRUCTION VEHICLES AND BUSES 103

Wrecker, 4-5/8" x 1-5/8" x 1-3/4", with cab that tilts back to show motor, Service Truck, 4" x 1-1/2" x 1-1/2", with opening tool chest and a miniature hammer, screwdriver and wrench, assorted color combinations, plastic wheels, Gilmark Merchandise Corp., USA, early 1950s. Suggested Retail $0.10.

Super Highway Fleet, card 12-3/4" x 2-3/8" x 3", trucks 3-3/4" to 4-3/8" long, with cabs that tilt back to show motor, assorted color combinations, plastic wheels, Gilmark Merchandise Corp., USA (No. 57), early 1950s. Suggested Retail $0.29.

Pumper Fire Truck, 6" x 1-3/4" x 2", red with painted chrome trim, rubber wheels, Hubley Manufacturing Co., USA, 1949 to early 1950s.

Fire Chief's Car, red with black and silver painted trim, rubber wheels, 1949 to mid 1950s, Hook and Ladder Truck, 6-3/4" x 2-1/4" x 2-3/8", with removable ladders, cab that tilts forward to show motor, red, rubber wheels (No. 324), Hubley Mfg. Co., USA, early to mid 1950s.

Road Roller, 5-3/4" x 2-1/2" x 2", green with silver painted trim, wooden rollers, Hubley Mfg. Co., USA (No. 315), early 1950s. Suggested Retail $0.49.

Dump Truck, 6-1/4" x 2-1/4" x 2-3/8", cab tilts forward to show motor and bed lifts to dump, green and red, rubber wheels, Hubley Mfg. Co., USA (No. 322), early 1950s. Suggested Retail $0.49.

Motor Express, 6-7/8" x 2-5/16" x 2-3/4", cab tilts forward to show motor and tailgate drops, red and green, rubber wheels, Hubley Mfg. Co., USA (No. 330), early 1950s. Suggested Retail $0.49.

Tow Truck, 6-1/4" x 2-1/4" x 2-3/8", cab tilts forward to show motor, with working winch, red, green and white, rubber wheels, Hubley Mfg. Co., USA, early 1950s. Suggested Retail $0.49.

Pickup Truck, 4-5/8" x 1-5/8" x 1-1/2", assorted colors, plastic wheels, Ideal Novelty and Toy Co., USA (No. TR-2), 1945-1947. Suggested Retail in 1945 $0.25, 1946-1947 $0.10. With Teardrop Trailer, 3-1/8" x 1-7/16" x 1-1/8" (No. TT-1), 1946-1947.

Beverage Truck, 4-1/8" x 1-1/2" x 1-1/8", assorted colors, plastic wheels, Ideal Novelty and Toy Co., USA (No. BT-10), 1947-1949. Suggested Retail $0.10.

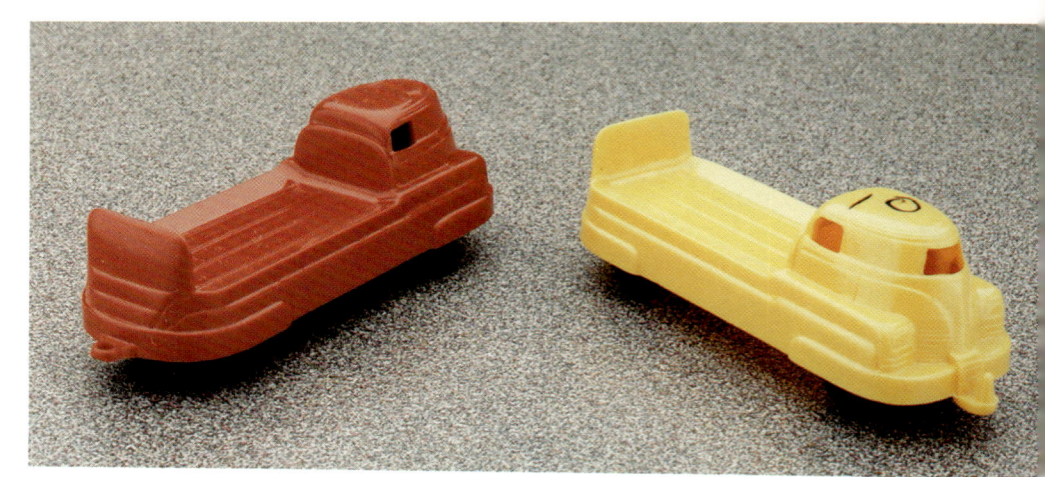

Pickup Truck, 4-1/4" x 1-1/2" x 1-1/4", assorted colors, plastic wheels, Ideal Novelty and Toy Co., USA (No. SP-10), 1947-1949. Suggested Retail $0.10. Reissued in 1950 as part of Road Building Set (No. RB-200).

Pickup Truck, 4-1/4" x 1-1/2" x 1-1/4", same as (No. SP-10) but with a plastic insert glued into the bed to make a gas tank that can be filled with water and a rumble seat, converted by the Deluxe Game Corp., USA for use in their service station playsets, 1947-1950.

Oil Truck, 4-1/8" x 1-1/2" x 1-1/8", assorted colors, plastic wheels, Ideal Novelty and Toy Co., USA (No. OT-10), 1947-1949. Suggested Retail $0.10. Reissued in 1950 as part of Road Building Set (No. RB-200).

Coal Truck, 5-1/2" x 2-1/2" x 1-3/4", when lever is pulled, body is raised and coal chute opens, assorted color combinations, plastic wheels, Ideal Novelty and Toy Co., USA (No. CT-50), 1947-1948. Suggested Retail $0.49. Not shown, Dump Truck, 5-1/2" x 2-1/2" x 1-3/4", same style cab (No. DT-50), 1947-1948.

Cement Mixer, 6-1/4" x 2-1/4" x 2-5/8", drum turns as truck moves and raises when lever is pulled, assorted color combinations, plastic wheels, Ideal Novelty and Toy Co., USA (No. CM-50), 1947-1950. Suggested Retail $0.49.

Tractor and Hayrake, 9" x 4-1/2" x 2-1/2", steering wheel turns and rake lifts up and down as hayrake is pulled, assorted color combinations, plastic wheels, Ideal Novelty and Toy Co., USA (No. HTR-100), 1947-1949. Suggested Retail $1.00.

TRUCKS, CONTRUCTION VEHICLES AND BUSES 107

Farm Implements; seeder 5-1/4" x 2-1/2" x 2", plastic wheels, mower 5-1/2" x 4-5/8" x 1-3/4", rubber wheels, assorted color combinations, Ideal Novelty and Toy Co., USA (No. STR-60), late 1940s. Suggested Retail $0.59. An uncataloged set.

Steam Shovel, 9" x 4-1/4" x 2-3/8", shovel cab revolves and digging action is operated by a lever, assorted color combinations, plastic wheels, Ideal Novelty and Toy Co., USA (No. SS-75), 1948-1951. Renumbered (No. 4888) in 1951. Suggested Retail $0.75.

Sanitation Truck, 6" x 2-1/4" x 2", sliding side doors permit loading and when lever is pulled, body is raised and backboard opens, assorted color combinations, plastic wheels, Ideal Novelty and Toy Co., USA, (No. STR-100), 1948-1949. Suggested Retail $1.00.

Seeder and Tractor, 9" x 4-1/2" x 2-1/2", steering wheel turns and two hoppers open and close alternately as seeder is pulled, assorted color combinations, plastic wheels, Ideal Nov-

108   PLASTIC TOYS

Mower and Tractor, 9" x 4-1/2" x 2-1/2", steering wheel turns and double row of teeth perform realistic cutting motion as mower is pulled, assorted color combinations, plastic and rubber wheels, Ideal Novelty and Toy Co., USA (No. MTR-100), 1948-1949. Suggested Retail $1.00.

Sanitation Truck, 6" x 2-1/4" x 2", sliding side doors permit loading and when lever is pulled, body is raised and backboard opens, assorted color combinations, plastic wheels, Ideal Novelty and Toy Co., USA (No. SAT-60), 1949-1950. Suggested Retail $0.59.

Panel Truck, 6-1/4" x 2-1/4" x 2-1/4", back doors open and close, assorted color combinations, rubber wheels, Ideal Novelty and Toy Co., USA (No. PT-40), 1949-1950, (No. 3085), 1951-1952. Suggested Retail $0.39.

Delivery Truck, 5" x 2-1/4" x 2-1/2", with driver, back doors open and close, side doors slide open, assorted color combinations, rubber wheels, Ideal Novelty and Toy Co., USA (No. TD-40), 1949-1950, (No. 3052), 1951-1952. Suggested Retail $0.39.

TRUCKS, CONTRUCTION VEHICLES AND BUSES

Dump Truck, 5-1/2" x 2-1/2" x 1-3/4", when lever is pulled, body is raised and backboard swings open, assorted color combinations, plastic wheels, Ideal Novelty and Toy Co., USA (No. DT-50), 1949, suggested retail $0.49, part of Road Building Set (No. RB-200), 1950 and (No. 4870), 1951. Suggested Retail $0.49. Not shown, Coal Truck, 5-1/2" x 2-1/2" x 1-3/4", same style cab (No. CT-50), 1949. Suggested Retail $0.49.

Carousel Truck, 6-1/2" x 3" x 3", carousel turns as truck is pulled, assorted color combinations, rubber wheels, Ideal Novelty and Toy Co., USA (No. MGR-75), 1949-1950. Suggested Retail $0.75. Not shown, Circus Seal Truck, 5-1/4" x 2-5/16" x 3", with seals on a rotating platform in a cage, same style cab (No. RS-70), 1950. Suggested Retail $0.75.

Road Roller, 4-1/2" x 2" x 2-1/2", with driver and steering roller, assorted color combinations, plastic wheels, Ideal Novelty and Toy Co., USA (No. RR-25), 1949-1950, suggested retail $0.25, part of Road Building Set (No. RB-200), 1950, (No. 4870), 1951.

Grocery Truck, 5" x 1-13/16" x 1-3/4", assorted colors with decals, plastic wheels, Ideal Novelty and Toy Co., USA (No. PA-10), 1950, (No. 3065), 1951-1952. Suggested Retail $0.10.

Milk Delivery Truck, 4-7/16" x 2" x 1-3/4", white with decals, plastic wheels, Ideal Novelty and Toy Co., USA (No. MD-10), 1950, (No. 3070), 1951-1952. Suggested Retail $0.10.

Moving Van, 4-3/8" x 1-11/16" x 1-7/8", assorted colors, plastic wheels, Ideal Novelty and Toy Co., USA (No. VA-10), 1950, (No. 3078), 1951. Suggested Retail $0.10.

Milk Truck, 4-3/4" x 2-1/4" x 1-1/4", with driver, back doors open and close, side doors slide open, white with decals, rubber wheels, Ideal Novelty and Toy Co., USA (No. MT-40), 1950, (No. 3067), 1951-1952. Suggested Retail $0.35.

Bungalow Bar Truck, 5-3/4" x 2-3/4" x 2-3/8", two side doors open, white and red plastic, rubber or plastic wheels, Ideal Novelty and Toy Co., USA (No. BBT-40), 1950, (No. 3044), 1951-1952. Suggested Retail $0.40.

TRUCKS, CONTRUCTION VEHICLES AND BUSES

Fix It Truck, 8-1/2" x 3-3/4" x 3-1/2", two opening tool boxes and side compartments with working jack, spare wheel, metal screwdriver, hammer and wrench, assorted color combinations, plastic wheels, Ideal Novelty and Toy Co., USA (No. FT-100), 1950, (No. 3059), 1951-1954. Suggested Retail $1.30.

Mechanical Fire Truck, 7" x 3" x 2-3/4", friction motor, with driver, assorted colors with painted details, Ideal Novelty and Toy Co., USA (No. MFE-60F), 1950, (No. 3332), 1951-1953. Suggested Retail $1.00.

Road Building Set, box 15" x 13-3/4" x 3-1/8", includes; tractor and plow, removable driver, cement mixer with square cab, dump truck with square cab, steam shovel with square cab, road roller, oil truck and pickup truck, assorted color combinations, plastic wheels, Ideal Novelty and Toy Co., USA (No. RB-200), 1950, (No. 4870), 1951. Suggested Retail $1.98.

Road Building Set, Ideal Novelty and Toy Co., USA (No. RB-200), 1950, (No. 4870), 1951.

Square Cab Cement Truck, 6-1/4" x 2-1/4" x 2-5/8", only available in Road Building Set, Ideal Novelty and Toy Co., USA (No. RB-200), 1950, (No. 4870), 1951.

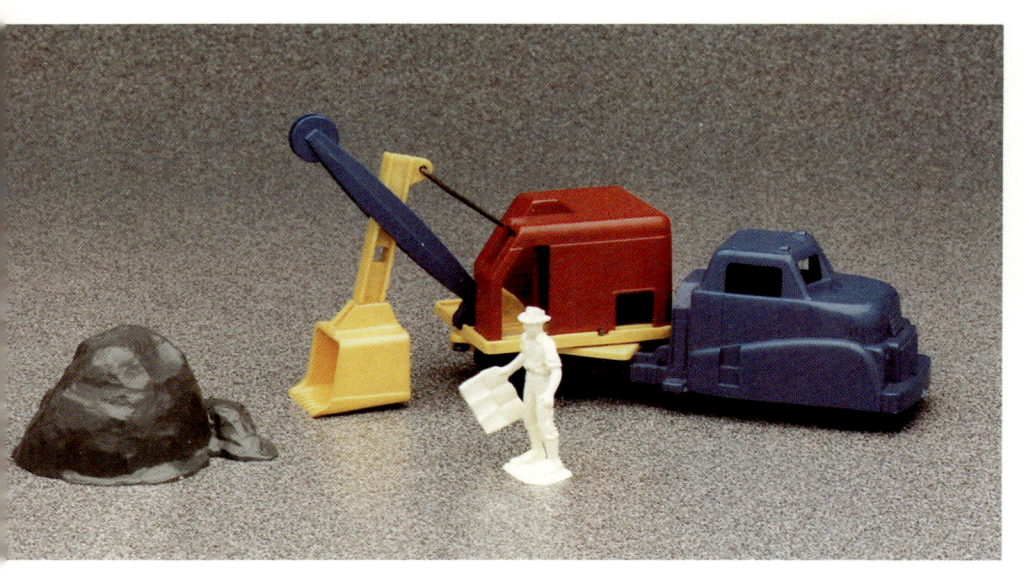

Square Cab Steam Shovel, 9" x 4-1/4" x 2-3/8", only available in Road Building Set, Ideal Novelty and Toy Co., USA (No. RB-200), 1950, (No. 4870), 1951.

TRUCKS, CONTRUCTION VEHICLES AND BUSES 113

Plough Wagon and Tractor, 12-5/8" x 3-3/16" x 3-3/8", plough on tractor moves up and down, detachable plow wagon has lever that dumps contents, assorted color combinations, plastic wheels, Ideal Novelty and Toy Co., USA (No. PWT-80), 1950, (No. 4790), 1951-1952. Suggested Retail $0.79.

Dump Truck, 5-3/4" x 2-1/4" x 2", when lever is pulled, body is raised and backboaard swings open, assorted color combinations, plastic wheels, Ideal Toy Corporation, USA (No. 3075), 1951-1954. Suggested Retail $0.20.

Beverage Truck, 4-1/2" x 1-1/2" x 1-1/4", assorted colors, plastic wheels, Ideal Toy Corporation, USA, part of assortment of trucks (No. 3025), 1951-1954. Suggested Retail $0.10.

Truck Assortment, consists of Gasoline Delivery Truck 3-3/4" long, Pick-up Truck 4"long, and Streamlined Fire Truck 3-3/4" long, assorted colors, plastic wheels, Ideal Toy Corporation, USA, 12 dozen per carton (No. 3057), 1951-1954. Suggested Retail $0.05.

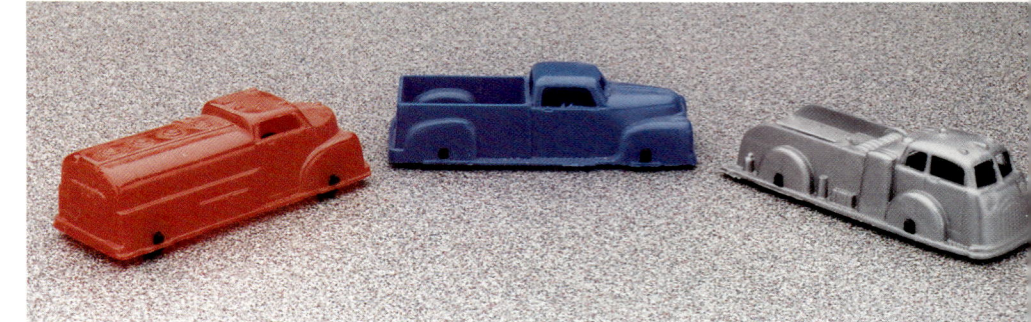

Paint Truck, 11-1/2" x 3-1/4" x 4-1/8", with opening cab and storage compartment doors, five plastic pails each containing a different water color, one empty pail for water, two paint brushes, two piece extension ladder, and one house ladder, assorted color combinations, rubber wheels, Ideal Toy Corporation, USA, (No. 3073), 1952-1954. Suggested Retail $1.70.

National Trailways Bus, 10" x 3-1/2" x 2-3/4", opening and closing passenger door and luggage compartment, side sign can be adjusted to show various U.S. cities, assorted color combinations, plastic wheels, Ideal Toy Corporation, USA, (No. 3093), 1952-1953. Suggested Retail $1.00.

Auto Delivery Van with three 3" long cars, 9-1/4" long, with 3" long cars and ramp that lowers, assorted color combinations, plastic wheels, Ideal Toy Corporation, USA, (No. 3055), 1952-1955. Suggested Retail $0.30.

TRUCKS, CONTRUCTION VEHICLES AND BUSES

Air Compressor Truck, Truck, 11" x 3-3/4" x 4-1/2", Trailer, 6" x 3" x 4-1/2", with opening doors on truck and trailer, simulated air compressor motor with wind-up knob on top of truck, pneumatic drill moves up and down as motor unwinds, trailer contains shovel, pick, lantern and three piece road block, assorted color combinations, rubber wheels, Ideal Toy Corporation, USA, (No. 3017), 1953-1954. Suggested Retail $2.80. Marketed by Kleeware in England.

Television Repair Truck, 8-3/8" x 3-7/8" x 3-1/8", with opening rear doors, two extension ladders, 2-1/2" tall antenna, television cabinet, picture tube and chassis, assorted colors, rubber wheels, Ideal Toy Corporation, USA, (No. 3019), 1953-1954. Suggested Retail $0.90.

Fire Engine, 8" x 2-1/4" x 2", with revolving two piece extension ladder, red, plastic wheels, Ideal Toy Corporation, USA, (No. 3037), 1953. Suggested Retail $0.35.

Emergency Truck, 12-3/4" x 5" x 5", with battery powered red light on top that flashes off and on as friction motor with siren races along, opening tool box with pick, shovel and sledge hammer; also included is a removable stretcher, fire extinguisher, two-piece extension ladder, three-piece road block and opening control box on side with microphone and Morse code clicker, blue with silver painted trim and lithographed tin interior of cab, rubber wheels with metal hub caps, Ideal Toy Corporation, USA, (No. 3301), 1954. Suggested Retail $4.00.

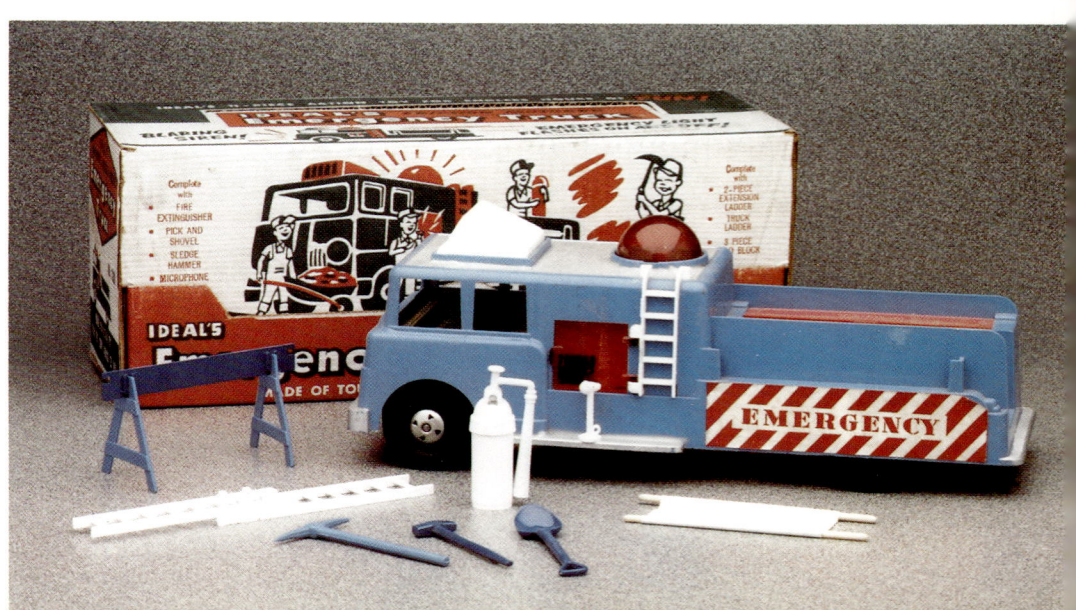

Blazer Fire Chief and Barking Dog, 6-3/4" x 2-1/2" x 4-1/2", with driver and dog, as fire engine is pushed, dog opens and shuts mouth and barks, red with painted details, plastic wheels, Ideal Toy Corporation, USA, (No. 4197), 1954-1955. Suggested Retail $0.90.

Old Smokey Comical Fire Engine, 12" x 5" x 4-3/4", with driver, when engine is pulled, front and rear sections raise and lower as bell rings, red with painted details, plastic wheels, Ideal Toy Corporation, USA, (No. 4198), 1954-1955. Suggested Retail $1.50.

Fire Engine with Aerial Ladder, 36-1/2" long x 6-1/2" wide, with rotating extension ladder that raises and extends to 44" by means of hand cranks and lever with realistic searchlight, and opening tool box, red plastic with aluminum ladder and bumper, die cast windshield and steering wheel and metal rack in rear, vinyl tires on chrome finish, plastic hubs, Ideal Toy Corporation, USA, (No. 3054), 1955-1957. Suggested Retail $12.00.

TRUCKS, CONTRUCTION VEHICLES AND BUSES

Tow Truck and Fix-It Car with Dented Fender, 20-1/2" x 6-1/2" x 6-1/2", metal towing hook can be raised and lowered by means of a crank, dented fender can be replaced with a new one and wheels on sedan can be removed and interchanged by means of a die-cast jack and wrench; working battery-powered searchlight works on or off the truck; three opening utility compartments and portable fire extinguisher; truck is plastic with aluminum bumper, side paneling, and tow crane, die-cast windshield, steering wheel and elevating mechanisms, wrench and ratchet-type jack, vinyl tires on plastic hubs, sedan is salmon color with painted black top, plastic wheels, Ideal Toy Corporation, USA, (No. 3001), 1955-1957. Suggested Retail $10.00.

Danger Patrol Truck, 12-3/4" x 5" x 5", with battery powered red light on top that flashes off and on as friction motor with siren races along, opening tool box with pick shovel and sledge hammer, also included is a removable stretcher, fire extinguisher, two-piece extension ladder, three-piece road block and opening control box on side with microphone and Morse code clicker, green with silver painted trim and lithographed tin interior, rubber wheels with metal hubcaps, Ideal Toy Corporation, USA, (No. 3300), 1955-1956. Suggested Retail $4.00.

Bus, 5-1/2" x 1-1/2" x 1-3/8", clear top shows detailed interior, assorted color combinations, plastic wheels, Irwin Corp., USA, late 1940s to early 1950s.

118 PLASTIC TOYS

Pickup Truck, 5-1/2" x 2-1/8" x 2", assorted colors including vacuum metalized, plastic wheels, Irwin Corp., USA, early 1950s.

Panel Truck, 6" x 2-3/8" x 2-1/4", assorted colors, plastic wheels, Irwin Corp., USA, early 1950s.

Bus, 7-1/2" x 1-5/8" x 1-7/8", with opening side door and destination sign that can be changed, assorted color combinations, plastic wheels, Keystone Mfg. Co., USA, early 1950s.

Express Truck, 4" x 1-3/8" x 1-3/8", assorted colors with gold hot stamping, plastic wheels, Kilgore Mfg. Co., USA, 1938 to late 1940s.

TRUCKS, CONTRUCTION VEHICLES AND BUSES

Bus, 4-3/8" x 1-3/8" x 1-3/8", assorted colors with gold hot stamping, plastic wheels, Kilgore Mfg. Co., USA, 1938 to late 1940s.

Stake Truck, 4-3/4" x 1-3/4" x 1-3/4", assorted colors, plastic wheels, Lapin Products Inc., USA (No. 285), early 1950s.

Ladder Fire Truck, 6" x 2" x 2-1/4", assorted colors, rubber wheels, Lido Toy Corp., USA, 1949 to early 1950s. Suggested Retail $0.10.

Pickup Truck, 6" x 1-5/8" x 2-1/4", Moving Van, 6" x 1-5/8" x 2-1/8", Gasoline Truck, 6" x 1-5/8" x 2-1/4", assorted colors, rubber wheels, Lido Toy Corp., USA, 1949 to early 1950s.

Traffic Lane, box 8" x 6-1/8" x 1", varied assortments of 3" long cars and/or trucks, assorted colors including vacuum metalized, rubber wheels, Lido Toy Corp., USA, 1949 to early 1950s. Suggested Retail $0.49.

Steam Shovel Construction Kit, box, no cementing required, assorted color combinations, rubber wheels, Lido Toy Corp., USA, early 1950s, originally manufactured and sold already assembled by California Moulders Inc., USA, 1949. Suggested Retail $0.79.

Truck, 6-1/2" x 2-13/16" x 2-1/2", assorted colors, plastic wheels, sold separately or as part of the Marx Trucking Terminal playset, Louis Marx and Co., USA, 1949 to early 1950s.

Coca Cola Delivery Truck, 11" x 3-1/2" x 4-1/2", with six removable miniature cases of Coca Cola, yellow with decals, rubber wheels, Louis Marx and Co., USA, with folding sides in 1949, without in 1950, cab and body changed in 1951. Suggested Retail $1.29.

Panel Truck and Delivery Truck, 4-1/8" x 1-1/2" x 1-1/2", wind-up with plastic gears and removable metal key, assorted color combinatins, rubber wheels, Louis Marx and Co., USA, late 1940s.

RCA Television Truck, 8-1/2" x 3-1/8" x 3-1/4", with a book of service orders, roof rack, extension ladder, and opening rear door, assorted colors with decals, wooden or rubber wheels, Louis Marx and Co., USA, early 1950s.

Pet Shop Delivery Truck, 11" x 3-1/2" x 4", with opening sides and six different vinyl dogs, assorted color combinations with red hot stamping, rubber wheels, Louis Marx and Co., USA, 1950 to mid 1950s. Suggested Retail $1.49.

Circus Truck, 11" x 3-1/2" x 5-3/8", with opening sides and the six following vinyl wild animals: tiger, leopard, alligator, buffalo, polar bear and cub, assorted color combinations with painted red raised letters, rubber or plastic wheels, Louis Marx and Co., USA (No. 2584), 1953 to mid 1950s.

Delivery Truck, 10" x 4" x 5", with opening rear doors, assorted color combinations with red hot stamping, plastic wheels, Louis Marx and Co., USA, early 1950s.

Delivery Truck, 10" x 4" x 5", Louis Marx and Co., USA, early 1950s.

FIX-ALL Wrecker Truck, 10-1/4" x 3-1/2" x 3-3/4", take apart wrecker with over thirty pieces, opening hood and four storage compartments in the rear, includes the following tools and equipment: spare wheel and tire, gas can, fire extinguisher, adjustable jack, hammer, open end wrench; box wrench, screwdriver, and adjustable wrench, assorted color combinations, plastic wheels, Louis Marx and Co., Inc., USA, 1953 to mid 1950s. Suggested Retail $1.49.

FIX-ALL Wrecker Truck, box 9-3/4" x 3-7/8" x 3-1/2", Louis Marx and Co., Inc., USA, 1953 to mid 1950s.

TRUCKS, CONTRUCTION VEHICLES AND BUSES

FIX-ALL Tractor, 9" x 4-3/4" x 6", take apart tractor with over forty pieces, removable driver, opening tool box, battery and radiator that can be filled with supplied water can, includes the following tools and equipment: gas can, fire extinguisher, adjustable jack, hammer, open end wrench, box wrench, screwdriver and adjustable wrench; assorted color combinations, rubber tires on plastic wheels, Louis Marx and Co., Inc., USA, 1953 to mid 1950s. Suggested Retail $2.49.

Take-A-Part Fire Truck, 4-1/2" x 1-7/16" x 1-1/2", red, plastic wheels, Louis Marx and Co., Inc., USA, early to mid 1950s.

Take-A-Part Wrecker, 4-7/8" x 1-3/8" x 1-7/8", silver, plastic wheels, Louis Marx and Co., Inc., USA, early to mid 1950s.

Ladder Fire Truck, 5-1/2" x 2" x 2", friction motor, red, plastic front wheels, rubber rear wheels, Louis Marx and Co., USA, early 1950s. Suggested Retail $0.59.

124 PLASTIC TOYS

Ice Cream Truck, 5-3/4" x 2" x 2-1/2", with friction motor, red, plastic front wheels and rubber rear wheels, Louis Marx and Co., USA, early 1950s. Suggested Retail $0.59.

Cement Truck, 5" x 1-3/4" x 1-7/8", Wrecker, 5" x 1 13/16" x 13/4", Sanitation Truck, 5" x 1-13/16" x 2", Pickup Truck, 5" x 1-13/16", 1-3/4", assorted colors, plastic wheels, Louis Marx and Co., USA, early 1950s. Suggested Retail $0.15 each.

Wrecker, 3-5/8" x 1-5/16" x 1-3/8", Flat Bed, 3-3/8" x 1-1/2" x 1-3/8", assorted colors, metal wheels, typical trucks found in Marx service station playsets, Louis Marx and Co., Inc., USA, early to mid 1950s.

Wrecker, 3-3/8" x 1-1/16" x 1-1/4", assorted colors, rubber wheels, Manoil Mfg. Co., USA, 1949 to early 1950s. Suggested Retail $0.05.

TRUCKS, CONTRUCTION VEHICLES AND BUSES   125

Pickup Truck, 3" x 1" x 1", assorted colors, rubber wheels, Manoil Mfg. Co., USA, 1949 to early 1950s. Suggested Retail $0.05.

Road Grader with Trailer, 5" x 1-1/8" x 1-1/2", assorted colors, rubber wheels, Manoil Mfg. Co., USA, 1949 to early 1950s. Suggested Retail $0.05.

Mystery Musical Ice Cream Truck, 11" x 4-1/2" x 4-3/4", with opening rear door, disappearing convertible sun canopy, music box that plays when the truck is pushed, includes a Popsicle premium catalog and one Popsicle wrapper, blue and red with metallic pressure-sensitive decals on sides, rubbber wheels, Mattel Inc., USA, 1955 to late 1950s. Suggested Retail $2.98.

Bus, 4" x 1-1/2" x 1-1/2", with wind-up motor and attached key, assorted colors, plastic wheels, Nosco Plastics, USA (No. 6335B.), 1948 to early 1950s. Suggested Retail $0.29.

Auto-Mite Wrecker, 3-3/8" x 1-1/2" x 1-3/8", with wind-up motor and attached key, assorted colors, plastic wheels, Nosco Plastics, USA (No. 6360), 1949 to early 1950s. Suggested Retail $0.19.

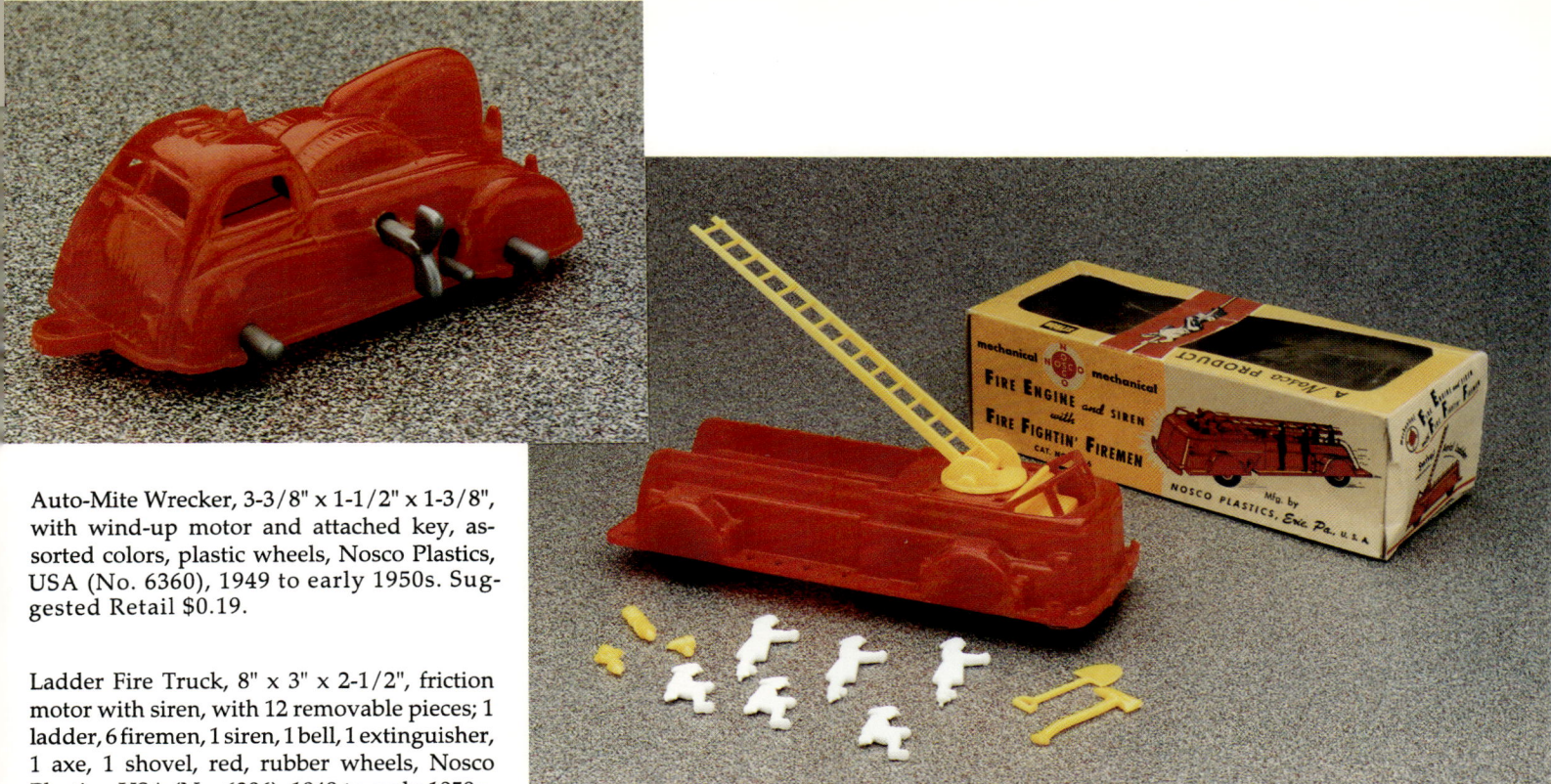

Ladder Fire Truck, 8" x 3" x 2-1/2", friction motor with siren, with 12 removable pieces; 1 ladder, 6 firemen, 1 siren, 1 bell, 1 extinguisher, 1 axe, 1 shovel, red, rubber wheels, Nosco Plastics, USA (No. 6386). 1949 to early 1950s.

Pokey Joe Ding Dong Fire Truck, 10" x 4-5/8" x 5-1/4", driver rings bell and firemen in back pump as toy is pushed or pulled, assorted color combinations, plastic wheels, Nosco Plastics, USA (No. 6430), 1951 through early 1950s.

TRUCKS, CONTRUCTION VEHICLES AND BUSES 127

Bus, 6" x 2-1/8" x 1-7/8", with "Hollywood Bus Lines" on its side, assorted colors, plastic wheels, Plas-Tex Corp., USA, 1948 to early 1950s.

Easter Bunny Truck, 4" x 1-3/8" x 1-3/4", when pushed, Easter Bunny pops up and down, assorted color combinations, plastic wheels, Plasticraft Mfg. Co., USA, early 1950s.

Mobile Crane, 10" x 3-1/2" x 5-1/4", with rotating cab and clam shell lift, red and yellow with or without painted chrome trim, rubber wheels, Processed Plastic Co., USA, 1952 to mid 1950s.

Dump Truck, 8-1/2" x 2-3/4" x 2-3/4", when lever is pulled bed lifts and dumps, assorted color combinations with white hot stamping, (No. 100) or olive drab (No. 120) and rubber wheels, Processed Plastic Co., USA (No. 100), 1948 to early 1950s. Suggested Retail $0.29.

Automatic Dump Truck, 10" x 3-3/4" x 3-1/2", spring loaded bed dumps automatically when lever is pushed, red and yellow with or without painted chrome trim, rubber wheels, Processed Plastic Co., USA (NO. 200), 1949 to early 1950s. Suggested Retail $0.49.

Tow Truck, 9-1/4" x 3" x 4", with metal tow crane and working winch, red with painted chrome trim (No. 300) or olive drab (No. 310) and rubber wheels, Processed Plastic Co., USA, 1951 through early 1950s.

Dump Truck, 7-3/4" x 2-3/4" x 2-3/4", bed lifts manually, red and yellow, rubber wheels, Processed Plastic Co., USA, early 1950s.

Bulldozer, 6-3/4" x 2-1/2" x 3", with scoop that lifts and dumps, red, yellow and blue, with concealed plastic wheels, Processed Plastic Co., USA, early 1950s.

Ladder Fire Truck, 6" x 1-7/8" x 2-1/4", with 2 steel ladders that connect, red, rubber wheels, Processed Plastic Co., USA, early 1950s.

TRUCKS, CONTRUCTION VEHICLES AND BUSES 129

International Pickup Truck, 9-1/2" x 3-1/2" x 3-5/8", yellow with decals, metal grill and front bumper, rubber tires on plastic hubs with metal hub caps, Product Miniature Co., USA, 1948.

Service Trucks, 4" x 1-1/4" x 1-1/2", assorted color combinations, plastic wheels, Pyro Plastics Corp., USA, early 1950s.

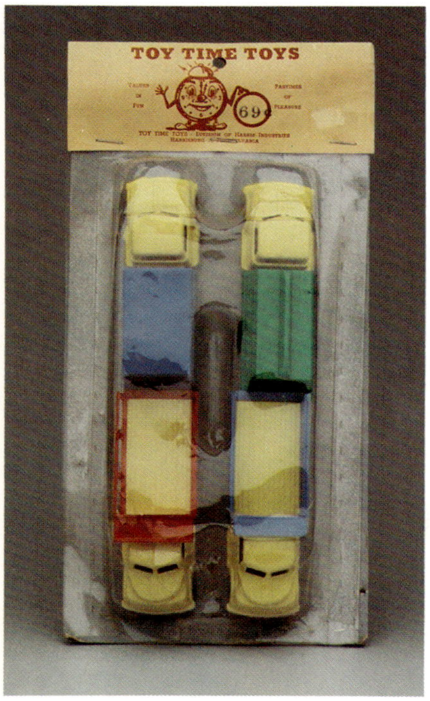

Service Trucks, blister card 11-1/4" x 5-1/2", trucks 4" x 1-1/4" x 1-1/2", assorted color combinations, plastic wheels, Pyro Plastics Corp., USA, early 1950s.

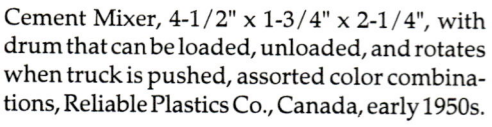

Cement Mixer, 4-1/2" x 1-3/4" x 2-1/4", with drum that can be loaded, unloaded, and rotates when truck is pushed, assorted color combinations, Reliable Plastics Co., Canada, early 1950s.

130   PLASTIC TOYS

Stake Truck, 6" x 2-3/8" x 2-1/8", with wind-up motor and detached metal key, driver, removable stakes (not shown) and opening doors, red, rubber tires on plastic wheels, Reliable Plastics Co., Canada, early 1950s.

Transport Truck, 10-3/4" x 3-1/4" x 2-1/2", cab with driver and opening doors, trailer with opening rear doors and one side door, assorted color combinations with decals on both sides of trailer, plastic wheels, Renwal Manufacturing Co., Inc. USA (No. 48), 1948-1954. Suggested Retail $0.79. As a construction kit (No. 223), 1954. Suggested Retail $0.79.

Coal Truck, 7-1/2" x 2-7/8" x 2-1/2", cab with driver and opening doors, load divider and unloading port with chute, body raises by means of a lever, assorted color combinations with a decal on one side of bed, plastic wheels, Renwal Manufacturing Co., Inc. USA (No. 46), 1948-1954. Suggested Retail $0.59. As a construction kit (No. 221), 1954. Suggested Retail $0.59. Not shown, Dump Truck, 7-1/2" x 2-3/4" x 2-1/2", cab with driver and opening doors, body raises by means of a lever and load slides out as gate swings open, assorted color combinations with a decal on one side of bed, plastic wheels, Renwal Manufacturing Co., Inc. USA (No. 50), 1948-1954. Suggested Retail $0.59. As a construction kit (No. 222), 1954. Suggested Retail $0.59.

Gasoline Truck, 7-3/4" x 2-1/4" x 2-1/4", cab with driver and opening doors, tank can be filled with water and drained through a faucet concealed behind an opening rear door, assorted color combinations with decals on both sides of tank, plastic wheels, Renwal Manufacturing Co., Inc. USA (No. 49), 1948-1953. Suggested Retail $0.59. As a construction kit (No. 224), 1954. Suggested Retail $0.59.

TRUCKS, CONTRUCTION VEHICLES AND BUSES 131

Cement Mixer Truck, 7-1/4" x 2-1/2" x 3-1/4", cab with driver and opening doors, drum can be filled and emptied when raised with a lever, drum revolves as truck is pushed, assorted color combinations with a decal on drum, plastic wheels, Renwal Manufacturing Co., Inc. USA (No. 56), 1948-1953. Suggested Retail $0.59. As a construction kit (No. 226), 1954. Suggested Retail $0.59.

Fire Truck, 7" x 2-1/2" x 2-1/8", with driver and two firemen, ladder swings to any position, hose winds and unwinds by means of a crank, red, plastic wheels, Renwal Manufacturing Co., Inc. USA (No. 57), 1948-1951. Suggested Retail $0.49. With painted chrome trim (No. 2057), 1952-1954. Suggested Retail $0.49. As a construction kit (No. 167), 1954. Suggested Retail $0.59.

Truck, 4-1/4" x 1-3/4" x 1-1/2", assorted colors, plastic wheels, Renwal Manufacturing Co., Inc. USA (No. 62), with chrome trim (No. 2621), 1948-1956. Suggested Retail $0.10.

Delivery Truck, 4-1/4" x 1-5/8" x 1-3/8", assorted colors, plastic wheels, Renwal Manufacturing Co., Inc. USA (No. 93), with painted chrome trim (No. 2093), 1949-1956. Suggested Retail $0.10.

Gasoline Truck, 4-1/4" x 1-5/8" x 1-3/8", assorted colors, plastic wheels, Renwal Manufacturing Co., Inc. USA (No. 94), with painted chrome trim (No. 2094), 1949-1956. Suggested Retail $0.10.

Auto Carrier, 13" x 3-1/4" x 2-7/8", cab with driver and opening doors; upper platform raises and lowers by means of a crank; rear ramp lowers; four vehicles included, usually two cars, one truck and one racer, assorted color combinations, later version had painted chrome trim, plastic wheels, Renwal Manufacturing Co., Inc. USA (No. 79), 1949. Suggested Retail $1.59.

Steam Shovel Truck, 8-1/8" x 3-3/8" x 9-5/8", cab with driver and opening doors, operator in shovel cab; shovel raises, lowers and dumps by means of a crank; assorted colors, plastic wheels, Renwal Manufacturing Co., Inc. USA (No. 86), 1949-1952. Suggested Retail $0.98. As a construction kit (No. 270), 1953-1954. Suggested Retail $1.29.

Auto Carrier, 13" x 3-1/4" x 2-7/8", cab with driver and opening doors, upper platform raises and lowers by means of a lever, rear ramp lowers , four vehicles included, usually two cars, one truck and one racer, assorted color combinations, painted chrome trim, plastic wheels, Renwal Manufacturing Co., Inc. USA (No. 79), 1950-1954. Suggested Retail $1.59.

TRUCKS, CONTRUCTION VEHICLES AND BUSES 133

Stake Wagon, 8-3/8" x 2-7/8" x 2-1/4", tongue hinged, front wheels turn, removable gates, assorted color combinations, plastic wheels, Renwal Manufacturing Co., Inc. USA (No. 99), 1950-1952. Suggested Retail $0.39. Also available as a Stake Truck, 7-3/4" x 2-7/8" x 2-1/4", cab with driver and opening doors, assorted color combinations, plastic wheels, (No. 101), 1950-1952. Suggested Retail $0.49.

Fire Engine, 4-1/4" x 1-3/4" x 1-3/8", with ladder that raises, assorted colors, plastic wheels, Renwal Mfg. Co., Inc., USA (No. 105), 1950-1954. Suggested Retail $0.10.

School Bus, 4-7/16" x 1-11/16" x 1-3/8", assorted colors, plastic wheels, Renwal Mfg. Co., Inc., USA (No. 123), 1950-1955. Suggested Retail $0.10.

City Bus, 4-7/16" x 1-11/16" x 1-3/8", assorted colors, plastic wheels, Renwal Mfg. Co., Inc., USA (No. 124), 1950-1955. Suggested Retail $0.10.

Pickup Truck, 3-3/16" x 1-1/16" x 1-1/16", (No. 149), 1950-1955; Gasoline Truck, 3-1/8" x 1-1/8" x 1-1/16", (No. 148), 1950-1955, assorted colors, plastic wheels, Renwal Manufacturing Co., Inc. USA. Suggested Retail $0.05 each.

Cement Mixer Truck, 9-7/8" x 5-1/4" x 6-1/2", cab with driver and opening doors, drum turns and tilts by means of two cranks, can be filled and emptied, assorted color combinations, plastic wheels, Renwal Manufacturing Co., Inc. USA (No. 131), 1951-1953. Suggested Retail $1.59.

Gasoline Truck, 12" x 3-3/8" x 3-1/4", cab with driver and opening doors, tank can be filled with water and emptied through plastic hose, two rear doors open to store hose, assorted color combinations, plastic wheels, Renwal Manufacturing Co., Inc. USA (No. 132), 1951-1953. Suggested Retail $1.59.

Hook and Ladder Truck, 15-3/4" x 2-5/16" x 3-1/2", cab with driver and opening doors, three firemen, movable extension ladder, three sections of removable ladder, assorted color combinations, plastic wheels, Renwal Manufacturing Co., Inc. USA (No. 126), 1950-1952. Suggested Retail $0.98. As a construction kit (No. 271), 1953-1954. Suggested Retail $1.29.

TRUCKS, CONTRUCTION VEHICLES AND BUSES

Fire House Set, 7" x 5" x 2-7/8", with two 3-1/4" long firetrucks (No. 145) and (No. 146), curved double doors that open, red with the words "Fire Dept." in white across the front, Renwal Manufacturing Co., Inc. USA (No. 263), 1955. Suggested Retail $1.00. Fire Trucks also sold individually as (No. 145) and (No. 146), 1950-1955, assorted colors, plastic wheels. Suggested Retail $0.05 each.

Fire Truck, 15" x 3-1/4" x 3-3/8", with two firemen, movable extension ladder, red, plastic wheels, (No. 178), 1953-1954. Suggested Retail $0.98. Fire Truck, 7-1/2" x 1-3/4" x 2-1/4", with two firemen, movable ladder, red, plastic wheels, (No. 179), 1953-1954. Suggested Retail $0.29. Renwal Manufacturing Co., Inc. USA.

Tractor, 5-1/4" x 2-3/4" x 3-3/4", assorted color combinations, plastic wheels, Renwal Manufacturing Co., Inc. USA (No. 186), 1953-1955. Suggested Retail $0.29.

Jr. Mechanic Gift Set, box 15-1/4" x 15-1/4" x 4-1/2", includes 10-1/2" long gardening service truck with ladder, wheelbarrow, lawnmower, flower pots, watering can, seeds and more, 11" long television service truck with ladder, TV, picture tube, chasis, tube tester and more, 10-1/2" long plumbing service truck with bathtub, vise, hacksaw, pipes, joints and more, assorted color combinations, plastic wheels, Revell Inc., USA, 1953 to mid 1950s. Trucks also sold separately.

136 PLASTIC TOYS

Caterpillar Motor Grader, 12" X 4" X 3-1/2", with steerable front wheels, adjustable blade and independent knee-action rear suspension, yellow, plastic wheels, Revell Toys, USA, 1952 to mid 1950s.

Caterpillar Tractor and Wagon, 12" x 3" x 2-3/4", with detachable wagon that hauls and dumps when lever is released, yellow, plastic wheels, Revell Toys, USA, 1952 to mid 1950s.

Ladder Fire Truck, 4-3/4" x 1-1/2" x 1-3/4", with driver and glued on ladder, assorted color combinations, plastic wheels, Ross Tool and Manufacturing Co., USA, early 1950s.

Dump Truck, 4-1/4" x 1-1/2" x 1-3/4", Moving Van, 4-5/8" x 1-5/8" x 2-1/8", Gasoline Truck, 4-5/8" x 1-5/8" x 1-5/8", assorted color combinations, plastic wheels, Ross Tool and Mfg. Co., USA, early 1950s. Suggested Retail $0.15.

TRUCKS, CONTRUCTION VEHICLES AND BUSES 137

Bubble Gum Truck, 7" x 2-1/8" x 3-1/2", rear gate swings down to allow access to bubble gum, assorted color combinations, Sidney A. Tarrson Co., USA, early 1950s.

Truck and Trailer, 9" x 2-1/8" x 1-1/4", detachable trailer, assorted colors, plastic wheels, Thomas Manufacturing Corp., USA (No. 16), 1947-1952. Truck alone (No. 18), 1947-1952.

Truck, 4-5/8" x 1-3/4" x 1-3/8", shown with earliest version, without Acme name on hood, step plates, fender skirts or rear bumper, Thomas Manufacturing Corp., USA (No. 18), 1947-1952.

Wrecker Truck, 5" x 2-1/8" x 2-3/8", metal working wench, boom and tow hook on chain, assorted colors, water decal on one side, plastic wheels, Thomas Manufacturing Corp., USA (No. 26), 1947-1951.

Texaco Gas Truck, 4" x 1-3/8" x 1-3/8", assorted colors, rubber wheels, Thomas Manufacturing Corp., USA (No. 40), 1947-1950. Suggested Retail $0.10.

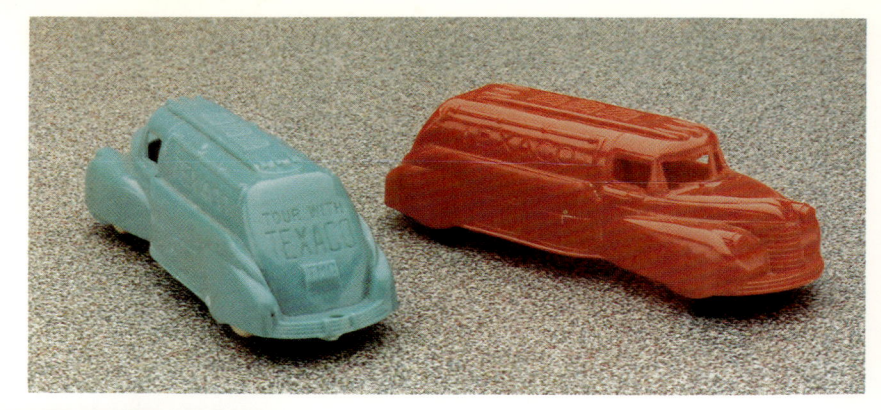

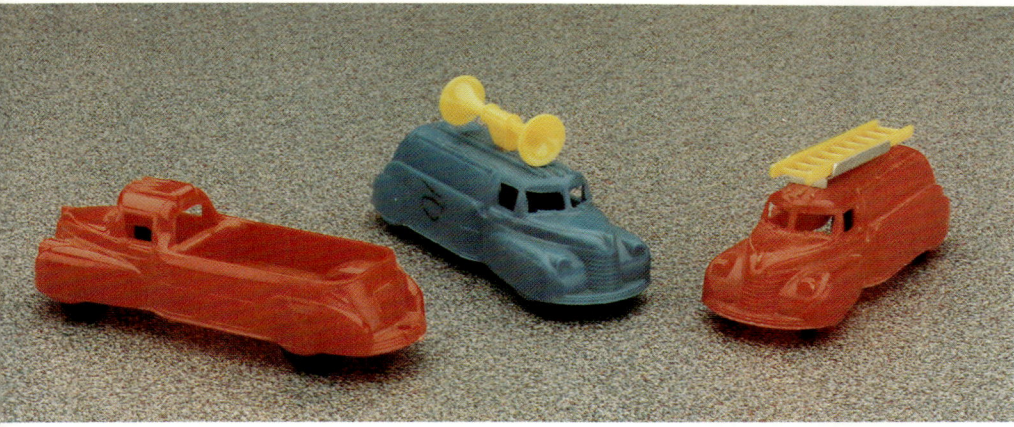

Delivery Truck, 4" x 1-3/8" x 1-3/8", (No. 41), 1947 to early 1950s, Sound Truck, 4" x 1-3/8" x 1-3/4", (No. 134), 1950 to mid 1950s, Repair Truck, 4" x 1-3/8" x 1-5/8", (No. 135), 1950 to mid 1950s, assorted colors and color combinations, rubber wheels, Thomas Manufacturing Corp., USA. Not shown, Tow Truck, (No. 139), 1950 to mid 1950s.

Merry-Go-Round Truck, 4-3/4" x 2-5/8" x 2-3/4", as truck rolls along, merry-go-round revolves, assorted color combinations, plastic wheels, Thomas Manufacturing Corp., USA (No. 74), 1949-1952.

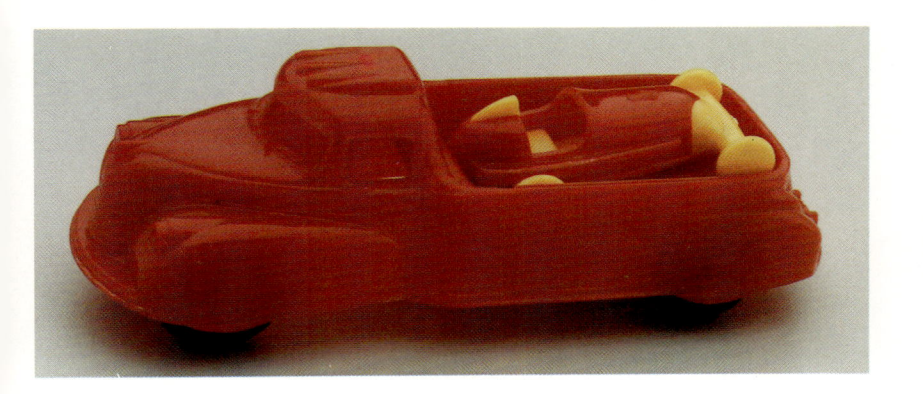

Truck and Racer, 4" x 1-3/8" x 1-3/8", with removable racing car, assorted colors and color combinations, rubber wheels on truck, fixed plastic wheels on racer, Thomas Manufacturing Corp., USA (No. 132), 1950 to mid 1950s. Racer alone (No. 141), 1950-1951.

TRUCKS, CONTRUCTION VEHICLES AND BUSES

Road Roller, 4-1/2" x 2-3/8" x 2-3/4", with driver and rubber band powered front roller that steers and turns, assorted colors, plastic rollers, Thomas Manufacturing Corp., USA (No. 126), 1951 to mid 1950s.

Bus, 5-1/2" x 1-1/4" x 1-3/4", with friction motor, "Inter-City Bus Lines" on its roof, green, rubber wheels, Manufacturer Unknown, USA, late 1940s to early 1950s.

Armored Car Bank, 5-1/2" x 2-1/4" x 2-7/8", with wind-up motor that is activated when coin is deposited through slot in roof, bronze, rubber wheels, Tri-Play Toys Inc., USA, mid 1950s.

Ladder Fire Truck, 3-1/2" x 1-1/4" x 1", red, plastic wheels, Manufacturer Unknown, USA, early 1950s.

140 PLASTIC TOYS

Ladder Fire Truck, 6-1/2" x 2-3/8" x 2-1/2", red, plastic wheels, Manufacturer Unknown, USA.

Ladder Truck, 9-1/2" x 3-5/8" x 3", missing ladders, red with painted chrome details, plastic wheels, Kay-Dee Plastics Inc., 1949 to early 1950s.

Wrecker, 3-1/2" x 1-3/8" x 1-1/4", yellow, plastic wheels, Manufacturer Unknown, USA, early 1950s.

Tow Truck, 5" x 1-3/4" x 2", assorted colors, plastic wheels, a Superior playset truck, possibly molded by Lido Toy Corp. for T. Cohen, USA.

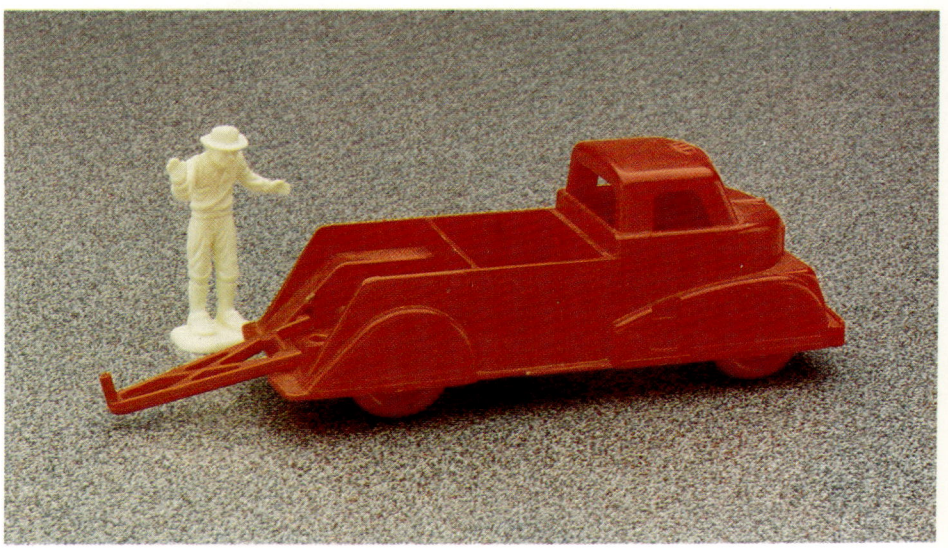

TRUCKS, CONTRUCTION VEHICLES AND BUSES 141

Delivery Truck, 5" x 1-3/4" x 1-3/4", with friction motor, green, plastic front wheels, rubber rear wheels, Manufacturer Unknown, USA, late 1940s to early 1950s.

Railway Express Truck, 4" x 2" x 2", with friction motor, green with decals, rubber tires on plastic wheels, Conway Company, late 1940s to early 1950s.

Sanitation Truck, 3-1/2" x 1-1/8" x 1-5/8", with friction motor, assorted color combinations, plastic front wheels, rubber rear wheels, Manufacturer Unknown, USA, early to mid 1950s.

Flat Bed Truck, 5-3/4" x 1-3/8" x 1-1/4", with pick and axe, assorted colors, plastic wheels, Manufacturer Unknown, USA, early 1950s.

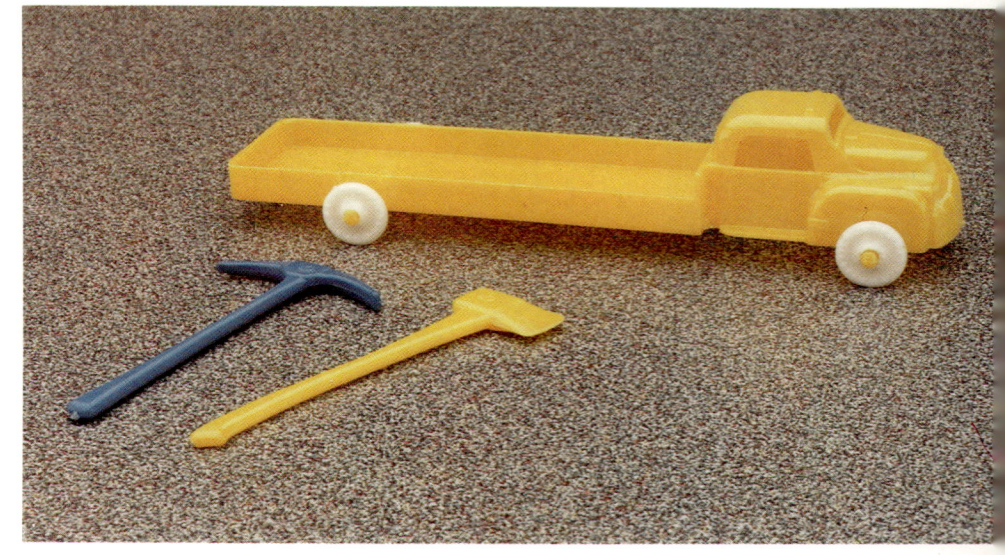

CHAPTER 10
BOATS

The first plastic toy boats were blow molded from sheets of cellulose nitrate, also known as celluloid. This method of manufacturing toys involves placing two 0.0015 inch thick sheets together in a heated mold and then blowing compressed air through a needle into the mold, forcing the heated material into the contours of the cavities. The combination of heat and pressure securely welds the two sheets together at points of contact creating a single unit. The mold is then cooled and the toys removed for trimming and hand decorating.

The 1890s saw both American and German toy makers exploiting this new manufacturing process. The United States, however, could not compete with Germany's experienced tool and die makers and its vast labor force composed mostly of poorly paid women.

Prior to the first World War, Germany accounted for most of the blow molded toys sold throughout the world. Shut off from the rest of the world during the war, Germany relinquished this position to a growing Japanese plastics industry.

With her vast resources of camphor from her island possession of Formosa and an extremely low labor rate, Japan soon became the world leader in the manufacturing and exporting of blow molded toys.

The threat of another World War and a growing concern over the highly inflammable nature of celluloid toys helped create the demand for the next generation of plastic boats.

The Tennessee Eastman Corporation lead the way in 1932, when it introduced Tenite, a cellulose acetate that was not only flame resistant, but also capable of being injection molded. At first, Tenite could only be compression molded. The technology necessary to do injection molding would not be developed until 1934 when the first modern injection molding machine would be invented in Germany.

The Dillon Beck Mfg. Co. of Hillside, New Jersey was the first American manufacturer to successfully mold and market toy boats using this new plastic.

In 1938, Edward W. Rowan, who would eventually become Dillon Beck Manufacturing Company's post-war president, was gainfully employed as an efficiency expert in the packaging industry. Then, fatefully, he was asked by a childhood friend, Daniel C. Dillon, Jr., if he would like to invest in the Sure Catch Lure Company of Hillside, New Jersey. The company was owned by George Beck, also a partner in the Eagle Tool and Machine Company of Newark, New Jersey.

The potential for injection molded fishing lures and tackle along with Beck's experience at the successful Eagle Tool and Machine Company, which had been making plastic molds since 1918, added up to a sound investment in the eyes of Rowan.

In 1939, Rowan, Dillon and two other investors purchased fifty percent of the Sure Catch Lure Company, each aquiring one eighth of the company. Shortly there-

Celluloid Battleship, 3 3/4" x 1 3/8" x 1 3/8", hand painted, Manufacturer Unknown, early 1900s.

after, Dillon was made president and Rowan vice President.

The lures were marketed under the clever slogan, "Wanna Catch the Limit" which appeared on all packaging. The future looked promising, despite the world's growing concern over the escalating hostilities in Europe and Asia.

This optimistic view was short lived, however. Hitler's armies invaded Poland on September 1, 1939. England and France responded quickly with declarations of war and their cry for munitions was soon heard in the United States.

On October 27, 1939, Congress responded to President Roosevelt's pleas by repealing the embargo on munitions that was part of the Neutrality Act of 1937. The U.S. response was too late for the Allies' immediate needs in Europe, however, and the following spring the British Expeditionary Force evacuated Dunkirk and Paris surrendered to the Germans. Congress again answered the President's call on July 2, 1940 by passing legislation authorizing unprecedented defense appropriations, totaling nearly nine billion dollars with another eight billion to follow by autumn.

With this kind of money pumped into the economy it began to thrive with full employment, stable prices due to government controls and the highest industrial output since the mid 1920s. While the future of manufacturing looked bright, the United Stated entry into the war seemed inevitable.

Should that happen, Dillon and Rowan knew, fishing tackle, with its metal components, would be eliminated as nonessential. There was a feeling at the time in both the plastics and toy industries that playthings helped preserve Young America's way of life and hopefully would not be classified as nonessential. With this in mind, Dillon and Rowan decided to shift the company's focus to toys. In 1940, they changed the name of the company to the Dillon Beck Manufacturing Company and introduced a bathtub fleet consisting of a ten cent submarine, a ten cent freighter and a fifteen cent cruiser at the 1941 Toy Fair. The new line was called "Wannatoy" after the earlier "Wanna Catch the Limit".

The demand for war related toys grew at a feverish pace. By July, 1941, Dillon reported that the company was using all available molding capacity to produce three ships and a well-designed model of the Bell P-39 Airacobra.

On December 7, 1941, Japan attacked Pearl Harbor and the next day the United States and Great Britain formally declared war on Japan. Three days later Germany and Italy declared war on the United States.

As expected, government restrictions on manufacturing soon followed as peaceful production ceased for all practical purposes and American industry geared up to meet the challenge.

Those companies molding plastic toys during the ensuing war years would not be allowed to purchase any new mold bases as tool steel was much too valuable to the war effort. They were also strictly forbidden from using any virgin molding material after the depletion of stocks on hand at the onset of such restrictions.

This eliminated most small shops who had been purchasing material on an as needed basis and who didn't have the luxury of having large inventories on hand.

The choice was simple; join the rush to secure scrap material from a dwindling number of sources or land a military contract with a guaranteed allotment of virgin material and a steady income.

By the end of 1942 the Dillon Beck Manufacturing Company had turned to 100% defense work, turning out molded parts for sextants, razors and aneometers.

Earlier in the year, the mold for the P-39 and an unfinished mold for a jeep that was to have been added to the line were sold to the Ideal Toy and Novelty Company of Long Island City, New York.

In 1943, Rowan met O.J. Sharpie, the sales manager of the Bergen Toy and Novelty Co. located in Rutherford, New Jersey. Sharpie, who had expressed an interest in having his own company, left Bergen in January, 1944 to start Plastic Toys Inc., in Cambridge, Ohio.

Sharpie's decision to move to Cambridge was greatly influenced by a deal he was able to put together with the plastics division of the Continental Can Company, also located in Cambridge. The deal secured an unlimited

Submarine, 4-1/2" x 1" x 1-3/8", assorted colors, Dillon Beck Manufacturing Company, USA, 1941 to 1950. Suggested Retail $0.05.

Victory Fleet, Freighter, 5-1/2" x 1" x 1", Battleship, 5-1/2" x 1" x 1-1/8", Aircraft Carrier, 5-5/8" x 1" x 3/4", assorted color combinations, Acme Plastics Mfg. Co., USA, Ideal Novelty and Toy Co., USA, 1942-1944.

amount of olive drab scrap plastic left over from Continental's military contracts. The color was perfect for Plastic Toys' new line on toy soldiers, which were direct copies of Bergen's except for their round, integral base.

Before moving to Cambridge, Sharpie had struck a deal with Rowan for the loan of the three boat molds, agreeing to pay Dillon Beck a royalty on each unit sold.

The molds for the cruiser and freighter were modified to require less plastic and less assembly time. The cruiser no longer had a gun and an airplane which had to be glued on separately. The freighter no longer had the fore and aft cargo cranes which were once molded as part of the hull and highly sucseptible to breakage. The submarine was not modified.

All three boats were molded using scrap plastic, and their quality is quite inferior to the original Dillon Beck versions.

Plastic Toys Inc. appears to have molded the boats from 1944 until 1950. The molds were never returned to Dillon Beck, who considered their designs outdated after the war.

Examples of the submarine can still be found today, identifiable by a large DB embossed on both sides of the hull. Original Dillon Beck cruisers and freighters are extremely hard to find and are considered quite rare. The modified Plastic Toys versions are almost as difficult to find.

The Dillon Beck Mfg. Co., which flourished after the war and into the early 1950s when Rowan left, would eventually just fade away.

Following close behind Dillon Beck was the Ideal Novelty and Toy Co. which introduced its "Victory Fleet" in 1942 as part of its newly formed plastic division.

The molds for the fleet which included a battleship, aircraft carrier and freighter were actually made in 1941 by Consolidated Molded Products Corporation of Scranton, Pennsylvania under the direction of their chief engineer, Islyn Thomas.

The three boats were the brainchild of Benjamin Shapiro, President and Owner of Acme Plastics Manufacturing Co. of New York, N.Y. which was really nothing more than a marketing company.

Shapiro, who had worked with Thomas since 1934 on a successful series of plastic toy filmstrip viewers, had originally intended to have Consolidated also do the molding. When Consolidated converted to 100% military work by the end of 1941, Shapiro was left high and dry without a molder or a source for plastic.

About the same time, Thomas was persuaded to leave Consolidated for a position as General Manager of the Ideal Novelty and Toy Co.. Thomas arranged for Ideal to mold the three boats for his friend, Shapiro, in exchange for an arrangement whereby both Ideal and Acme would share in the marketing of the boats. Acme would sell only individual boats and Ideal would sell only sets of boats packaged as a "Victory Fleet". This arrangement provided Ideal with its first injection molded toys for boys and explains why these boats are found without any molder's mark or any other form of identification on them.

These three boats were produced from 1942 through the end of 1944 when Shapiro pulled the molds to go into business with Thomas who left Ideal at that time. Judging by the number of examples remaining, they were eminently more popular or better engineered than those produced by Dillon Beck and later, Plastic Toys. The fact that they were also marketed as part of the Ideal line certainly helped in their success.

Ideal had stockpiled large quantities of the Acme Plastics boats, and was able to supply most of its customers through the end of 1945.

In 1946, Ideal replaced its "Victory Fleet" with four new cellulose acetate boats, a submarine, aircraft carrier, freighter and transport.

These boats appear to have been produced in molds provided by the Reliable Toy Co. Ltd. of Canada which had a history of sharing molds with Ideal. Only offered in 1946 and 1947, examples of the four are very scarce today.

Meanwhile, Thomas re-worked Shapiro's three boat molds adding new detail and reducing the height of the masts which broke off easily on the original versions.

These revised boats along with three airplanes, a helicopter, and a jeep made up Thomas Manufacturing Corporation's first offerings as a toy company in 1945.

The big news for 1946 was the introduction of four

BOATS 145

entirely new types of boats by the Dillon Beck Mfg. Company. These were the first injection molded PT boat and three different pleasure craft. All four were originally molded in cellulose acetate and the pleasure boats were later molded in polystyrene.

The PT was only made for a few years and is difficult to find. The three pleasure craft were molded into the late 1950s and are rather easy to come by.

A campaign by manufacturers to extend the toy buying season into spring and summer relied heavily on promoting such items as toy boats.

By the mid 1950s, the Thomas Mfg. Co. and the Ideal Toy Co. had emerged as the premier manufacturers of toy boats in the United States.

Each printed impressive catalogs of their boat lines, the likes of which would have made any arm-chair admiral envious.

While Ideal and Thomas were jockeying for the Number One position, Benjamin and Henry Hirsch were about to invent the best selling plastic toy boat of all time. In 1953, the Hirsch brothers, owners of a small cosmetic lab, discovered that the gas (carbon dioxide), created when baking powder comes in contact with water, could be used to raise a submerged object. Delighted with their discovery, they made a crude model of a submarine that could dive and surface and presented it to a major cereal company in nearby Battle Creek, Michigan.

The rest is history. The sub was introduced in 1954 for twenty-five cents and one box top. It took three shifts to keep up with the orders for the four and one half inch long polystyrene working model of the atomic submarine U.S.S. Nautilus. By May of that year, over one million subs had been produced and work was started on a two and one eighth inch polyethylene version that was given away by the tens of millions. The success of the submarine transformed Hirsch Laboratories, the cosmetic company, into a premium toy company almost overnight.

Their diving U.S. Navy Frogmen and working PT boat, both offered in two sizes, are two more examples of a long line of hits that would establish Hirsch Laboratories as the undisputed leader in premiums.

By the late 1950s, the inexpensive plastic toy boat was going the way of the inexpensive plastic toy automobile. Foreign competition had begun to take its toll. While several U.S. manufacturers, lead by Remco Industries of Newark, New Jersey, tried to counter foreign sales with large and impressive offerings in the 1960s, most simply sailed off into the sunset, never to be heard from again.

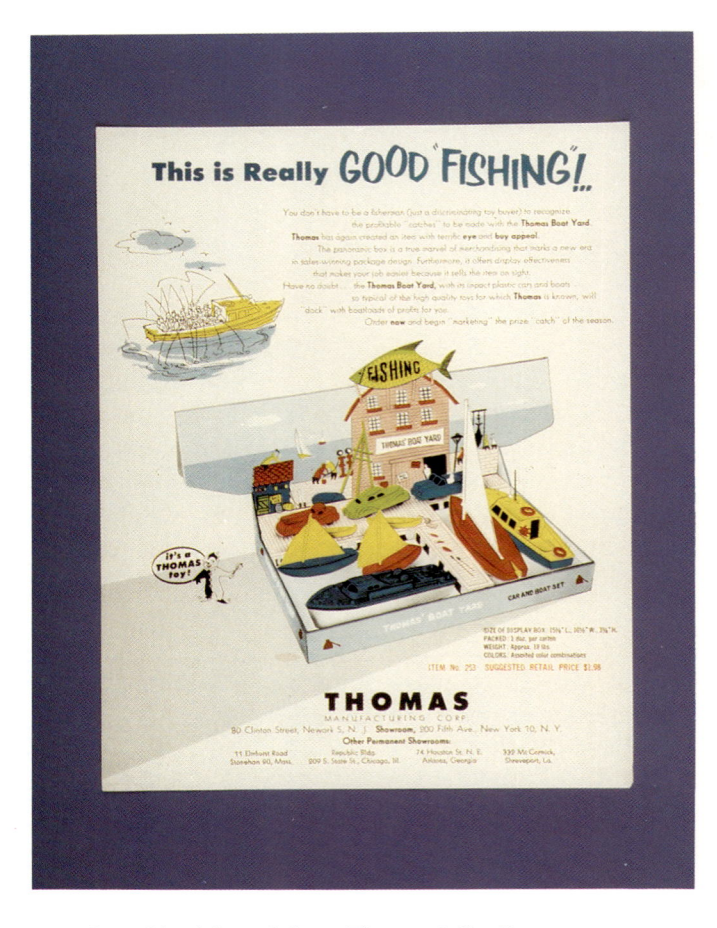

Boat Yard Set ad sheet, Thomas Mfg. Corp., USA, 1954.

U.S.S. Nautilus send-in version, 4-1/2" long, giveaway version 2" long; U.S. Navy Frogmen send-in version (set), 3-1/2" tall, giveaway version 2-1/4" tall; PT Boat send-in version, 5 1/4" long, giveaway version, 2-1/8" long, Hirsch Laboratories, USA, 1954 to late 1950s.

Torpedo Attack, box 15" x 10" x 2", submarine, 7" x 2" x 1-3/4", fires 3" long solid aluminum torpedo at battleship 10" x 3" x 2-1/2", which explodes if target on side is hit; green, Allstate Engineering Service, USA, 1947, forerunner and possible inspiration for Thomas Manufacturing Corporation's Torpedo Attack of the mid 1950s.

Fireboat, 3" x 3-1/4" x 9", rolls on land and floats in water; press the smokestack to shoot a strong jet of water twelve feet or more! No filling necessary when boat is in water, red and white, concealed plastic wheels, Amerline, USA, 1952 to mid 1950s. Suggested Retail $0.98.

Tugboat, 3-1/2" x 1-1/2" x 1-1/2", assorted color combinations, Banner Plastics Corp., USA, late 1940s to mid 1950s, railroad barge and crane barge, not shown. Suggested Retail $0.10.

Aircraft Carrier, 4" long, Battleship, 4" long, Banner Plastics Corp., USA, 1946 to mid 1950s.

BOATS 147

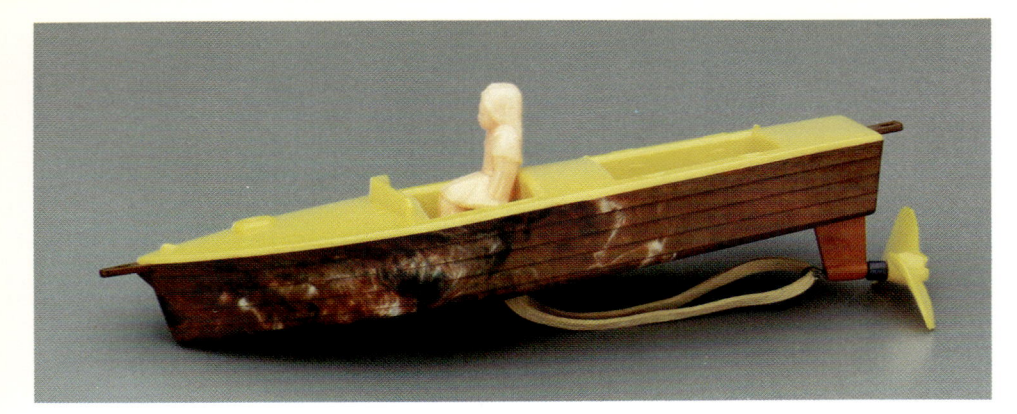

Speedboat, 7" x 2-1/8" x 1-3/4", rubber band powered with girl driver, assorted color combinations, Best Plastics Corp., USA, late 1940s to early 1950s, example shown missing rudder.

Submarine, 4-1/2" x 1" x 1-3/8", assorted colors, Dillon Beck Mfg. Co., USA, 1941-1950. Suggested Retail $0.05.

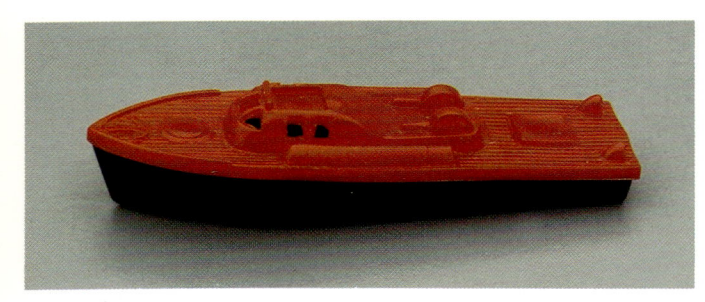

PT Boat, 4" x 1-1/4" x 1", assorted color combinations, Dillon Beck Mfg. Co., USA, (No. B2), 1946-1948. Suggested Retail $0.10.

Pleasure Boats, 4" x 1-1/4" x 1", includes three different speed boats, assorted color combinations, Dillon Beck Mfg. Co., USA (No. B-3), 1946 to mid 1950s. Suggested Retail $0.10.

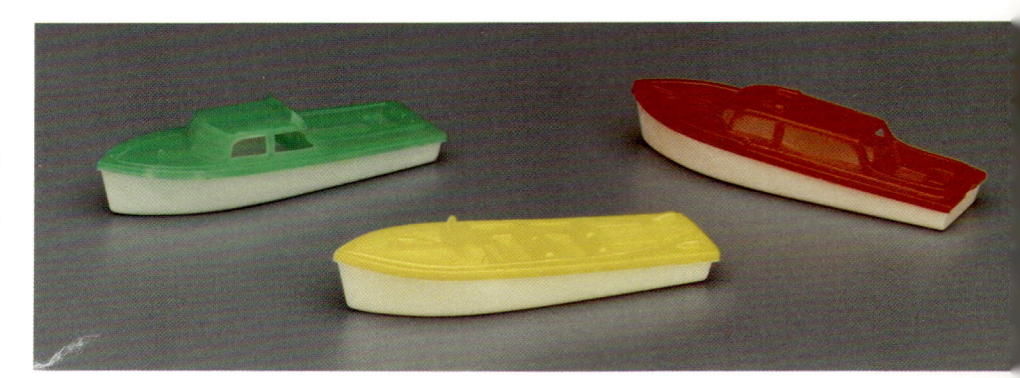

Victory Fleet, consists of a 5-1/2" x 1" x 1-1/8" battleship, a 5-5/8" x 1" x 3/4" aircraft carrier and a 5-1/2" x 1" x 1" freighter, Ideal Novelty and Toy Co., USA. (No. N 59) a four piece set, and (No. N 100) an eight piece set, 1942-1944. Suggested Retail in 1942, $1.00 and $0.59.

148   PLASTIC TOYS

Ocean Liner, 12-1/8" x 3-1/8" x 2-1/2", assorted color combinations, Ideal Novelty and Toy Co., USA (No. OS-75), 1950, (No. 4725), 1951-1954. Suggested Retail $0.60.

Speed Boat, 5-7/8" x 1-5/8" x 1-1/8", assorted color combinations, Ideal Novelty and Toy Co., USA (No. SB-40), 1947-1950, (No. 4740), 1951-1952. Suggested Retail $0.15.

Cabin Cruiser, 5-7/8" x 1-5/8" x 1-3/4", assorted color combinations, Ideal Novelty and Toy Co., USA (No. CC-40), 1947-1950, (No. 4717), 1951-1952. Suggested Retail $0.15.

Speed Boat with Outboard Motor, 13-7/16" x 5-1/2" x 4", detachable outboard motor with windup motor, assorted color combinations, Ideal Novelty and Toy Co., USA (No. MSB-150), 1950, (No. 4030), 1951-1955. Suggested Retail $2.00.

Mechanical L'll Abner Canoe, 12" x 3" x 3-1/2", wind-up motor, jug of "Joy Juice" serves as wind-up key, when wound "Lonesome Ploecat" paddles, assorted colors with painted details, Ideal Toy Corporation, USA (No. 4191), 1951-1952. Suggested Retail $2.00.

River Steamer, 5" x 2-1/4" x 2", assorted color combinations, Ideal Toy Corporation, USA (No. 4710), 1951-1954. Suggested Retail $0.15.

Houseboat, 5" x 2" x 1-1/2", assorted color combinations, Ideal Toy Corporation, USA (No. 4709), 1951-1954. Suggested Retail $0.15.

Great Lakes Ore Boat, 12" x 3-1/2" x 4-1/2", with working derrick and two pieces of cargo, rolls on land and floats in water, assorted color combinations, concealed plastic wheels, Ideal Toy Corporation, USA, (No. 4731), 1952-1953. Suggested Retail $1.30.

P.T. Boat, 8" x 2-1/2" x 2-3/4", with removable hatch, two rotating machine guns, assorted color combinations, Ideal Toy Corporation, USA, (No. 4350), 1951-1954. Suggested Retail $0.60.

150 PLASTIC TOYS

Pirate Ship, 13" x 6-1/2" x 11-1/2", with six white vinyl pirates, wooden masts, vinyl sails, crows nest, skull and crossbone flag, anchor, "walking plank", yellow dinghy with two oars and a working cannon with cannon balls; rolls on land and floats in water, assorted color combinations of red and blue, concealed plastic wheels, Ideal Toy Corporation, USA, (No. 4037), 1953-1958. Suggested Retail $3.00.

Pirate Ship detail, Ideal Toy Corporation, USA (No. 4037), 1953-1958.

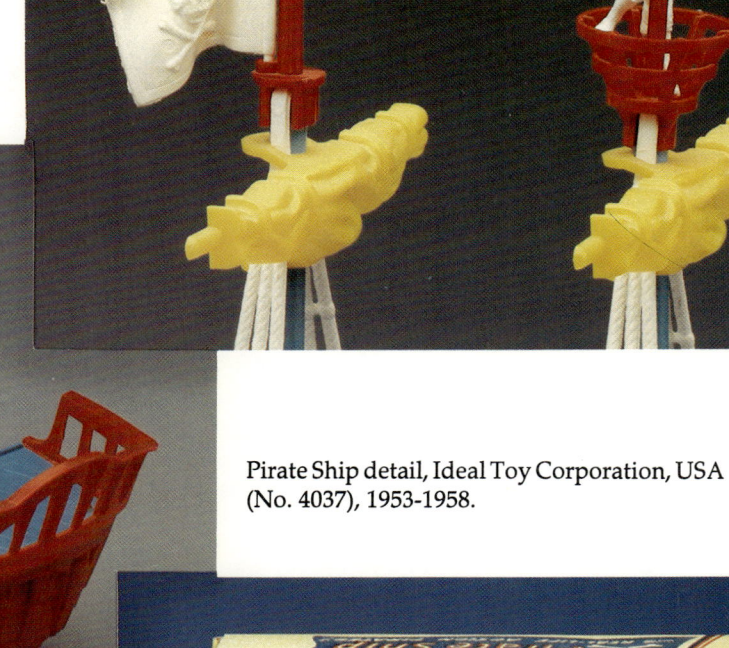

Pirate Ship detail, Ideal Toy Corporation, USA (No. 4037), 1953-1958.

Pirate Ship, box 14" x 6-3/4" x 9-1/2", Ideal Toy Corporation, USA (No. 4037), 1953-1958.

BOATS 151

Torpedo-Shooting Submarine, 12" x 3" x 4", with two plastic torpedoes which can be fired, opening hatch allows sub to be filled with water so it can submerge or float, assorted color combinations, Ideal Toy Corporation, USA, (No. 4705), 1953-1954. Suggested Retail $1.00. Reissued as Sea Wolf Atomic Sub, gray and black with blue camouflage painted color, (No. 4719), 1957, as U.S.S. Nautilus Atomic Submarine, (No. 4895-9), 1959-1962, as Atomic Submarine, 1963, and as Polaris Sub, silver hull with red top (No. 31419), 1964-1969.

Luxury Speed Cruiser, 14-1/4" x 5-5/8", wind-up motor with attached key, assorted color combinations, Ideal Toy Corporation, USA, (No. 4025), 1953-1955. Suggested Retail $3.00.

Boat Assortment, consists of; racing boat 4" long, inboard speed boat 4" long, inboard cabin cruiser 4" long and luxury speed boat 4" long, assorted color combinations, packed one gross per carton, Ideal Toy Corporation, USA, (No. 4033), 1953-1954. Suggested Retail $0.05.

Take-Apart Harbour Launch, 14" long x 5" wide, twenty-five detailed parts can be disassembled by removing screws and then reassembled again and again; accessories include fire extinguisher, two life preservers, two Marlin spikes, ladder, grappling hook, paper flag on wooden pole, die cast anchor with lever to raise and lower it, white hull with mahogany color deck, red and blue accessories, Ideal Toy Corporation, USA, (No. 4713), 1954-1957. Suggested Retail $3.00.

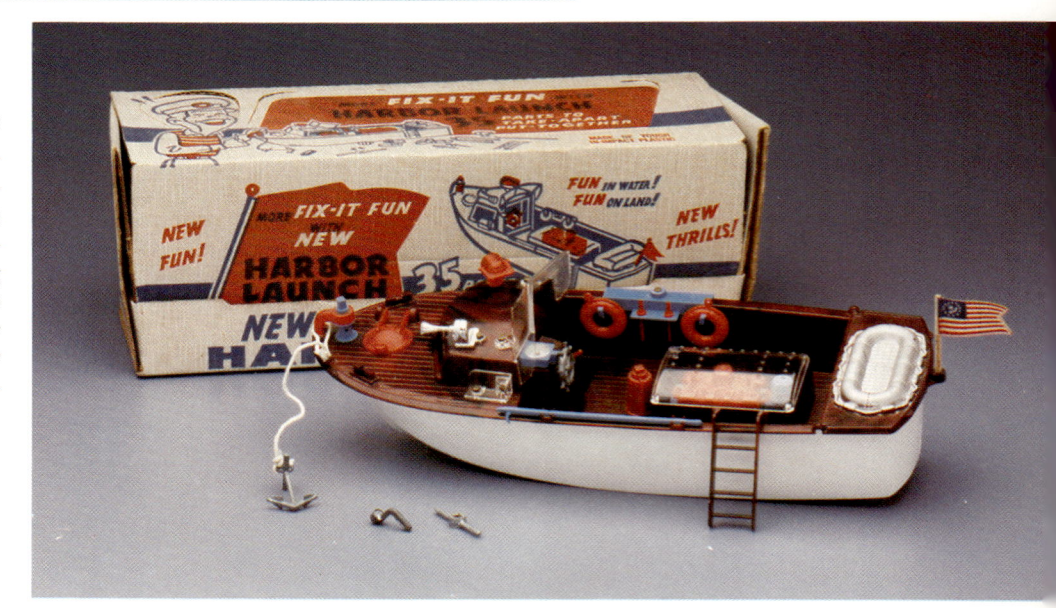

Sparking Mechanical Torpedo Boat, 11-1/4" long, wind-up motor with captain at wheel and rear gunner in sparking double gun turret, replaceable flint, attached key, assorted color combinations of white and gray with decal on one side of hull, Ideal Toy Corporation, USA, (No. 4024), 1954-1957. Suggested Retail $2.00. Reissued as PT Boat 109, blue and gray, (No. 3144-3), 1964-1965.

Inertia Motor Racing Boat, 13" long, when crank on deck is turned, powerful inertia motor zooms boat along for long ride, with driver and adjustable rider, assorted color combinations with painted accents, Ideal Toy Corporation, USA, (No. 4022), 1954-1957, stabilizing fins added in 1955. Suggested Retail $2.00.

Inertia Motor Racing Boat (SLO-MOSHUN VI), with stabilizing fins, Ideal Toy Corporation, USA (No. 4022), 1955-1957.

BOATS 153

U.S. Navy Destroyer, 15" long, inspired by Columbia Pictures, "The Caine Mutiny", with three rotating guns, three firing torpedo tubes with five polyethylene torpedos, two deck racks to drop eight wooden depth charges and a "K" gun to launch two more wooden depth charges, rolls on land and floats in water, assorted color combinations of red, white and blue, concealed plastic wheels, Ideal Toy Corporation, USA, (No. 4711), 1954-1957. Suggested Retail $2.00. Reissued as Nuclear Destroyer, same colors, (No. 3142-7), 1964-1965.

Walt Disney's Mickey Mouse Old Fashioned Sailing Vessel, 13" x 6-1/2" x 11-1/2", with working cannon and cannon balls, gang plank, polyethylene dinghy, oars, anchor and sails, vinyl rigging, wooden masts, plastic crow's nest and flag pole without flag; the following eight yellow polyethylene characters: Mickey Mouse, Minnie Mouse, Donald Duck, Daisy Duck, Huey, Dewey, Louie and Pluto, rolls on land and floats in water, assorted bright color combinations, concealed plastic wheels, Ideal Toy Corporation, USA, (No. 4038), 1956-1957.

Walt Disney's Mickey Mouse Old Fashioned Sailing Vessel Crew, includes Donald Duck, Daisy Duck, Pluto, Huey, Minnie Mouse, Mickey Mouse, Dewey and Louie, Ideal Toy Corporation, USA, (No. 4038), 1956-1957.

Mickey Mouse Raft, 5" x 3-3/4", Sail 5-1/2" high, Mickey, Donald, and Pluto ride their simulated log raft along with their gold chest and water keg, polyethylene figures vary, assorted color combinations, Ideal Toy Corporation, USA, (No. 4039), 1956-1957.

Automatic Reversing Tugboat, 9-1/2" long, wind-up motor with smokestack acting as key, when wound, tug goes forward, stops, goes backwards again and again, assorted color combinations, Ideal Toy Corporation, USA, (No. 4021), 1956-1957.

Mechanical Speedboat with Aquaplane Rider, Boat 12" long, Aquaplane Rider 3-1/2" high, wind-up motor concealed under two opening doors on deck, start/stop lever in cockpit, mahogany color with metalized deck accessories, Ideal Toy Corporation, USA, (No. 4026), 1956-1957.

Water Pumping Mechanical Fireboat, 15" long, when crank is turned, siren screams and two rotating fire-guns pump a steady stream of water, with working life boat, anchor, and two life preservers, rolls on land and floats in water, red and white with metalized and painted deck accessories, Ideal Toy Corporation, USA, (No. 4714), 1956-1959. Suggested Retail $4.98.

BOATS 155

Mechanical Harbor Police Boat, 14" x 5" x 5", with inertia motor and loud siren, two opening storage compartments in rear contain; rifle, shotgun, tommy gun, binoculars, and die cast winch wrench, also included are a fire extinguisher, life preserver, two boat hooks and rotating machine gun, assorted color combinations of dark and light green with red painted waterline and yellow accessories, Ideal Toy Corporation, USA, (No. 4027), 1956-1957.

Power Speedboat, 21" long, with wind-up motor, assorted color combinations, Ideal Toy Corporation, USA, (No. 4040), 1957-1958.

Power Speedboat Cockpit detail, Ideal Toy Corporation, USA (No. 4040), 1957-1958.

156 PLASTIC TOYS

Polar Queen, 13-1/2" x 3-1/2", with working harpoon gun and two polyethylene harpoons, helicopter and landing platform, anchor, two polyethylene penguins, polar bear and removable icicles, large plastic whale, rolls on land and floats in water, assorted color combinations, concealed plastic wheels, Ideal Toy Corporation, USA, (No. 4716), 1957-1959. Suggested Retail $2.00.

Salvage Boat with Diver, 14-3/4" x 5-1/4" x 7-3/4", crane swings out to lower or retrieve diver, air pump and selector valve can be made to lower diver, keep him below water and raise him, depending on position of selector valve, includes two removable life preservers and ladder, rolls on land and floats in water, assorted color combinations, concealed plastic wheels, Ideal Toy Corporation, USA, (No. 4715), 1957. Five frogmen figures were added and the name changed to "Treasure Hunter with Diver" (No. 4729), 1958-1959. Suggested Retail $5.00.

Troop Transport, 12-1/2" x 3-1/4" x 4-3/4", replica of "Victory Ship" troop transport, gray blue camouflage paint scheme, Ideal Toy Corporation, USA (No. 3143-5), 1964-1965.

BOATS 157

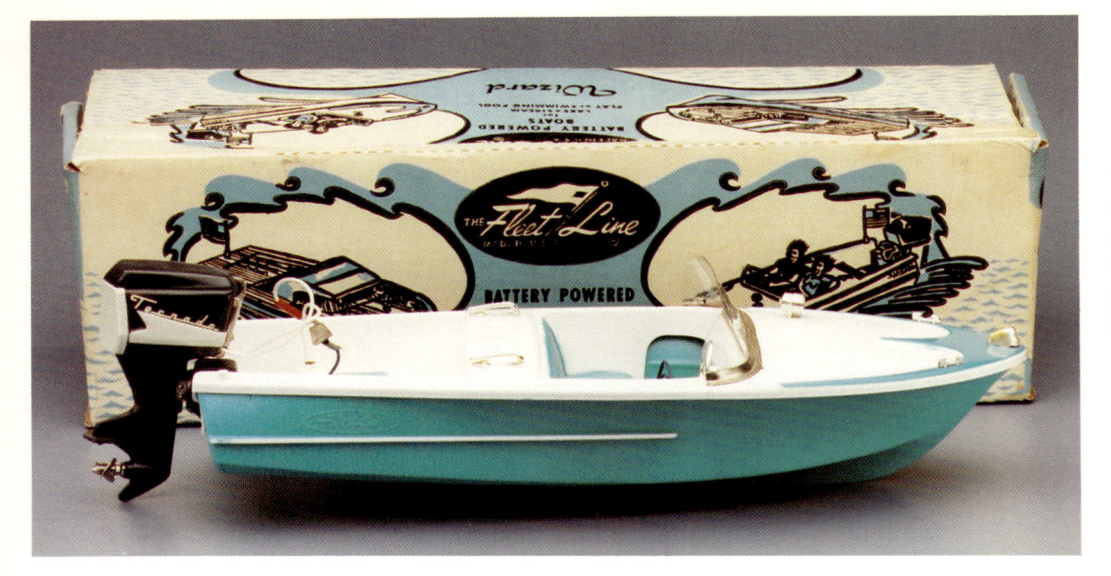

Wizard Speed Boat, 12-1/2" x 5-1/4" x 3-1/2", with separate battery powered outboard motor, assorted color combinations of blue and white with vacuum metalized trim, K & O Models Inc., USA, mid to late 1950s. Suggested Retail $4.98.

Cabin Cruiser, 6" x 2" x 1-1/2", assorted color combinations, Knickerbocker Plastic Co., Inc., USA, 1946 to early 1950s.

PT Boat, 6" x 2" x 1-1/2", assorted colors, Knickerbocker Plastic Co., Inc., USA, 1946 to early 1950s (press-in revolving guns missing from turrets).

Cruiser, 2-1/2" long, Ocean Liner, 1-1/4" long, assorted colors, Lido Toy Corp., USA, late 1940s to early 1950s.

River Queen, 8-1/2" x 4-1/4" x 5", wind-up motor with attached key, removable smokestacks and ramp, rolls on land and floats in water, assorted color combinations of red, yellow and green, plastic wheels, Louis Marx and Co., USA, early 1950s.

Indians in a Mighty Canoe, 13" x 2-1/2" x 3", with three removable polyethylene Indians, Multiple Products Corporation, USA, early 1950s.

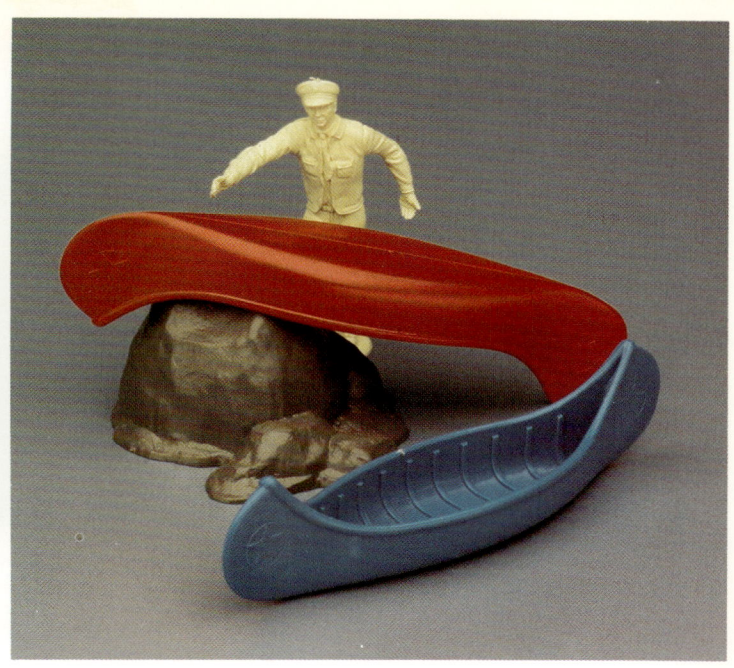

Canoe, 5-1/2" x 1-3/8" x 1-1/4", a Marx playset boat, assorted colors, Louis Marx and Co., USA, mid 1950s.

Cap'n Noah and Polly, 8" x 4-1/2" x 4", windup motor with Cap'n Noah acting as the key, works in water and on land, assorted color combinations, Revell Toys, USA, early 1950s.

Pirates of Treasure Island, box 12-1/2" x 4-1/2" x 2", standing figures 3-1/2" tall, assorted metallic colors, Plasticraft, USA, early to mid 1950s.

BOATS 159

Ferry Boat, 7-1/2" x 2-3/4" x 3", rolls on land and floats in water, with four 2-1/4" long sedans and/or coupes with fixed wheels, assorted color combinations, including olive drab with white hot stamping, plastic wheels, Pyro Plastics Corporation, USA, late 1940s to early 1950s.

River Dredge, 7-1/4" x 2-3/4" x 3-1/4", rolls on land and floats in water, with revolving cab and shovel that lifts by means of crank and chain, assorted color combinations, including olive drab with white hot stamping, plastic wheels, Pyro Plastics Corporation, USA, early 1950s.

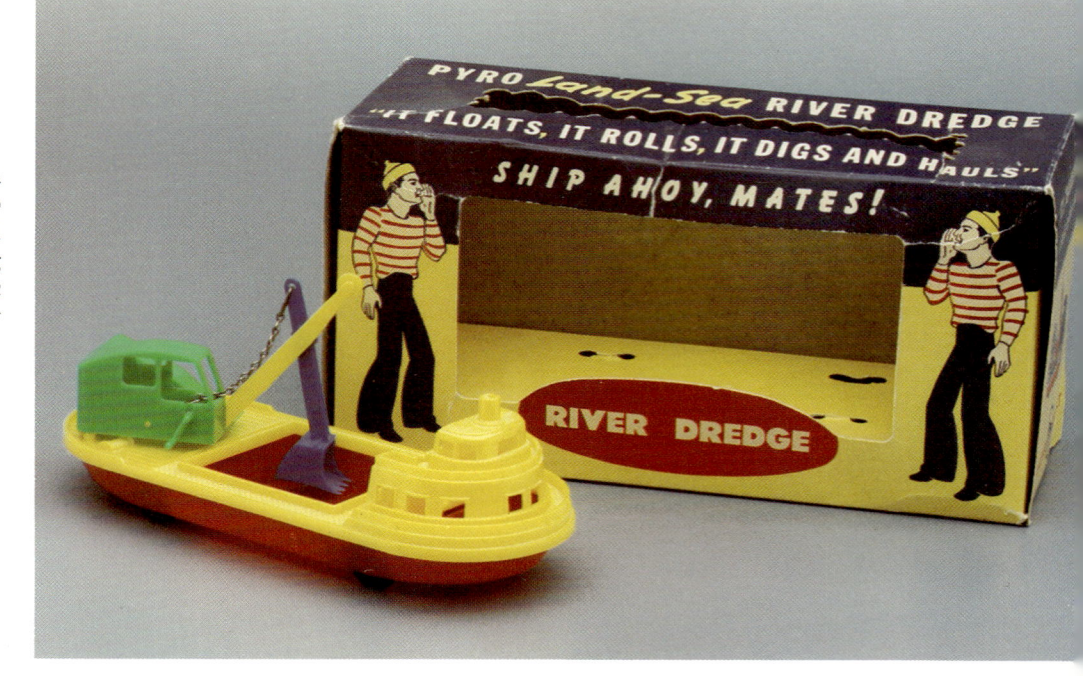

Noah's Ark, 7" x 3-3/8" x 3", rolls on land and floats in water, with the following eight animals; pig, dog, cow, elephant, camel, horse, giraffe and lion (not in pairs), assorted color combinations, plastic wheels, Pyro Plastics Corporation, USA, early 1950s.

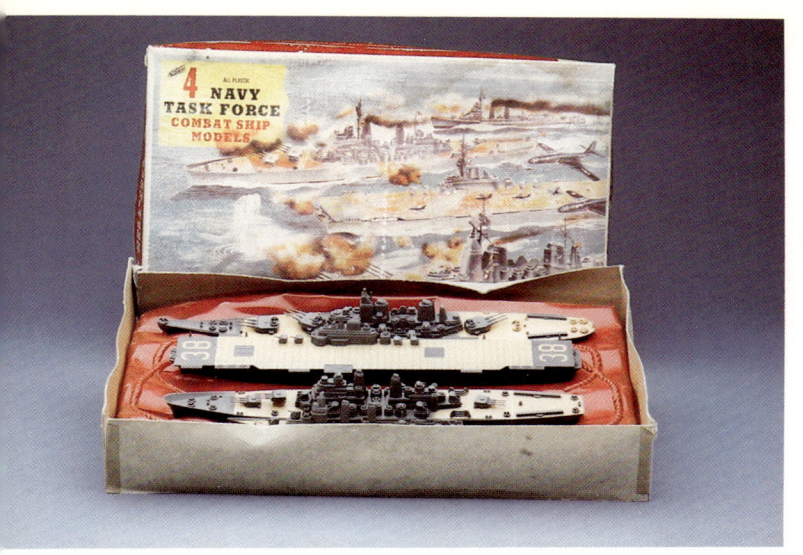

Four Navy Task Force Combat Ship Models, box 18" x 11-3/4" x 3", four piece set includes; the 9-3/4" long destroyer U.S.S. Allen M. Sumner, the 14-1/2" long cruiser U.S.S. Chicago, the 14-1/2" long aircraft carrier U.S.S. Shangri-La and the 16" long battleship Missouri, rolls on land and floats in water, held in a red vacuum formed tray, gray with painted tan decks, concealed plastic wheels, Pyro Plastics Corporation, USA (No. 283), early 1950s. Suggested Retail $3.98.

U.S. Navy Destroyer - U.S.S. Allen M. Sumner, 3-3/4" x 1-1/4" x 2", rolls on land and floats in water, with four revolving guns, gray with tan painted deck, concealed plastic wheels, Pyro Plastics Corporation, USA, early 1950s.

U.S. Navy Cruiser - U.S.S. Chicago, 14-1/2" x 1-7/8" x 2-3/4", rolls on land and floats in water, with three revolving guns, gray with tan painted deck, concealed plastic wheels, Pyro Plastics Corporation, USA, early 1950s.

U.S. Navy Aircraft Carrier - U.S.S. Shangri-La, 14-1/2" x 2-1/2" x 2-3/4", rolls on land and floats in water, with four revolving guns, gray with tan painted deck, concealed plastic wheels, Pyro Plastics Corporation, USA, early 1950s.

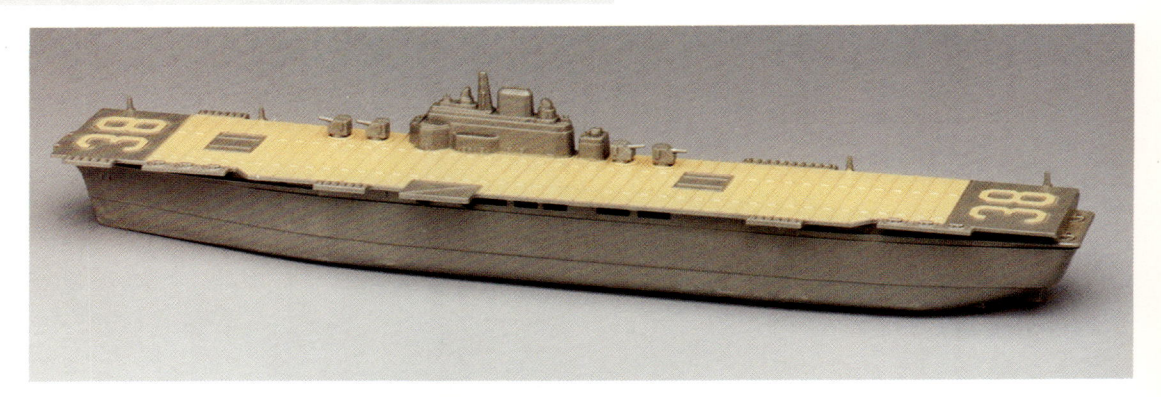

U.S. Navy Battleship - U.S.S. Missouri, 16" x 2-1/8" x 3", rolls on land and floats in water, with three revolving guns, gray with tan painted deck, concealed plastic wheels, Pyro Plastics Corporation, USA, early 1950s.

BOATS 161

Fire Fighter Boat, 10-1/2" x 3-1/4" x 4-1/4", floats or rolls, tank in hull holds water which is pumped through hose when water tower is pushed up and down, assorted color combinations, concealed plastic wheels, Renwal Manufacturing Co., Inc. USA (No. 156), 1952-1955. Suggested Retail $1.49.

Boat Construction Kit, 3-1/4" to 4-7/8" long, seventy pieces build a fleet of seven boats, includes (No. 136) canoe, (No. 137) rowboat, (No. 138) ocean liner, (No. 139) freighter, (No. 140) ferry, (No. 141) motor cruiser, and (No. 142) tug; cement included, assorted color combinations, Renwal Manufacturing Co., Inc. USA (No. 231), 1954. Suggested Retail $0.98.

Canoe, 4-7/8" x 1-5/8" x 1-3/8", floats or rolls, with boy and girl, two paddles, assorted color combinations, concealed plastic wheels, Renwal Manufacturing Co., Inc. USA (No. 136), 1951-1956. Suggested Retail $0.15.

Rowboat, 4-1/2" x 1-7/8" x 1-3/8", floats or rolls, with boy and girl, two oars, assorted color combinations, concealed plastic wheels, Renwal Manufacturing Co., Inc. USA (No. 137), 1951-1956. Suggested Retail $0.15.

Ocean Liner, 4-1/2" x 1-3/8" x 1-5/16", floats and rolls, assorted color combinations, concealed plastic wheels, Renwal Manufacturing Co., Inc. USA (No. 138), 1951-1956. Suggested Retail $0.15.

Freighter, 4-1/4" x 1-3/8" x 2-3/16", floats and rolls, assorted color combinations, concealed plastic wheels, Renwal Manufacturing Co., Inc. USA (No. 139), 1951-1956. Suggested Retail $0.15.

Ferry, 3-1/4" x 1-3/4" x 1-9/16", floats and rolls, assorted color combinations, concealed plastic wheels, Renwal Manufacturing Co., Inc. USA (No. 140), 1951-1956. Suggested Retail $0.15.

Motor Cruiser, 4" x 1-5/8" x 1-9/16", floats and rolls, with boy and girl, assorted color combinations, concealed plastic wheels, Renwal Manufacturing Co., Inc. USA (No. 141), 1951-1956. Suggested Retail $0.15.

Tug, 4" x 1-3/4" x 2", floats and rolls, assorted color combinations, concealed plastic wheels, Renwal Manufacturing Co., Inc. USA (No. 142), 1951-1956. Suggested Retail $0.15.

BOATS 163

Ocean Freighter, 10" x 2-1/2" x 3-5/8", floats and rolls, assorted color combinations, concealed plastic wheels, Renwal Manufacturing Co., Inc. USA (No. 171), 1953-1956. Suggested Retail $0.59.

Ocean Liner, 10" x 2-1/2" x 3-1/8", floats and rolls, assorted color combinations, concealed plastic wheels, Renwal Manufacturing Co., Inc. USA (No. 172), 1953-1956. Suggested Retail $0.59.

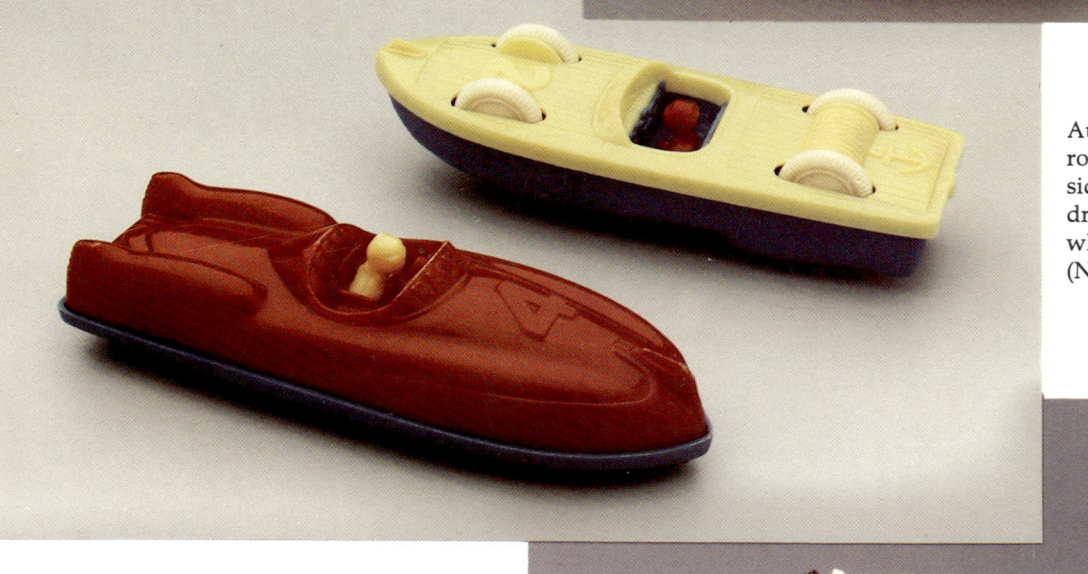

Auto-Boat, 6-1/2" x 2-1/4" x 2", floats and rolls, on one side it's a speed boat, on the other side it's a race car, each side with its own driver, assorted color combinations, plastic wheels, Renwal Manufacturing Co., Inc. USA (No. 168), 1953-1955. Suggested Retail $0.39.

Torpedo Firing Patrol Boat, 12" x 3-1/2" x 5", floats and rolls, magazine holds six torpedoes which are fired one at a time by spring action levers, four gun turrets revolve, assorted color combinations, plastic wheels, Renwal Manufacturing Co., Inc. USA (No. 217), 1954-1955. Suggested Retail $1.49.

164 PLASTIC TOYS

Viking Ship, 17-1/4" x 5-3/4" x 13", floats and rolls, Vikings row as ship is pushed or pulled across the floor, catapult throws projectiles, four provided, red, concealed plastic wheels, Renwal Manufacturing Co., Inc. USA (No. 245), 1955-1956. Suggested Retail $4.00.

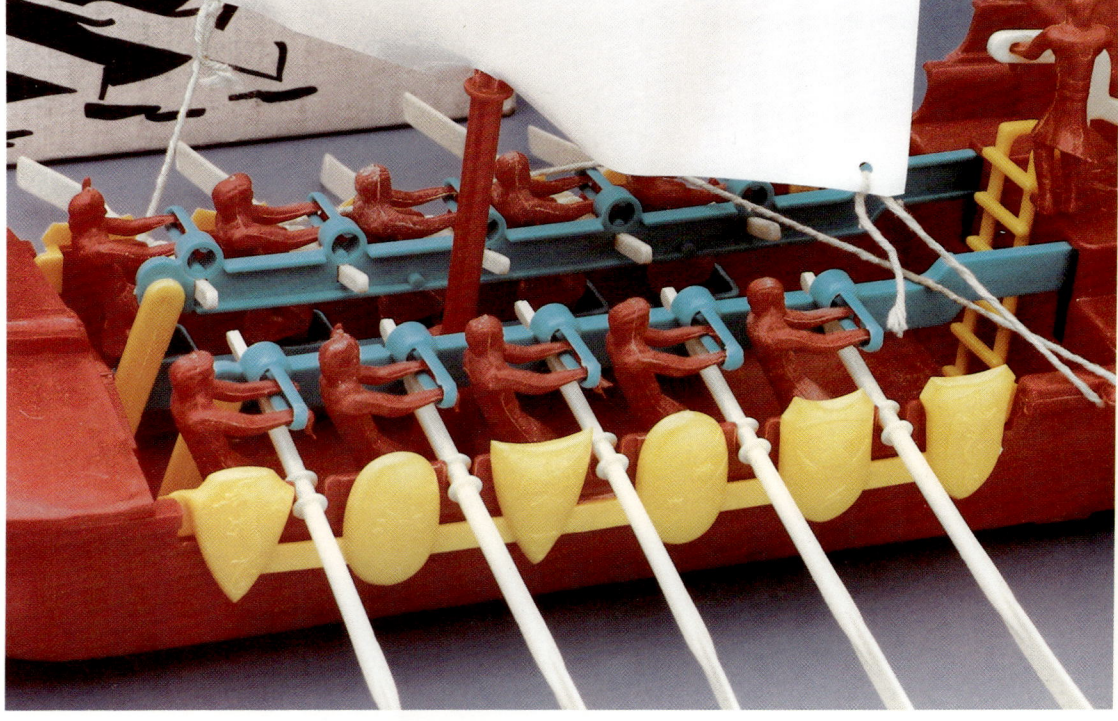

Viking Ship Detail, Renwal Mfg. Co., Inc., USA (No. 245), 1955-1956.

Super Catapult Plane Carrier, 12-1/4" x 3" x 3", with five jet planes and a rubber band powered launcher, assorted color combinations of red, yellow and blue with No. 38 decals on deck, rubber wheels, Saunders Tool and Die Co., USA (No. 38), 1951 to mid 1950s, Suggested Retail $1.50.

BOATS 165

5-1/2" Boat Assortment, 5-1/2" x 1" x 1-1/4", includes Queen Mary, Battleship, Aircraft Carrier and Freighter (not shown), assorted color combinations, Thomas Manufacturing Corp., USA (No. 15), 1945 to late 1950s. Suggested Retail $0.10. (The Queen Mary was added in 1946).

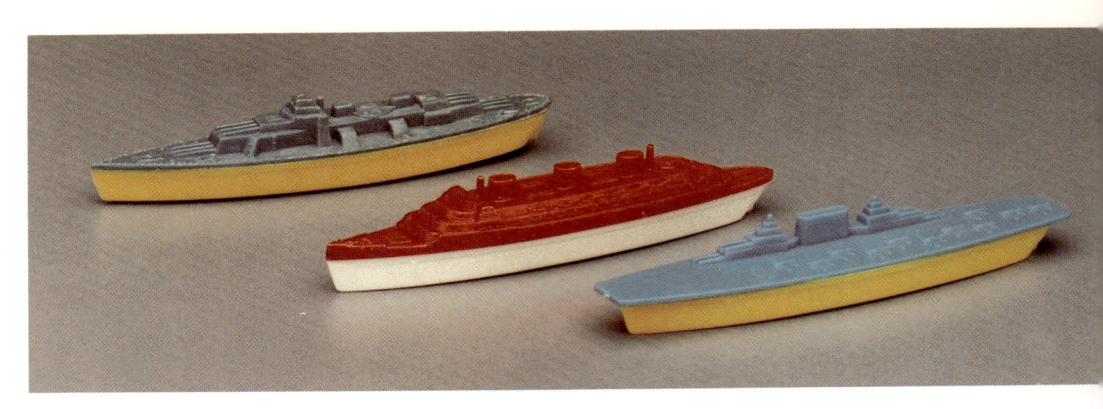

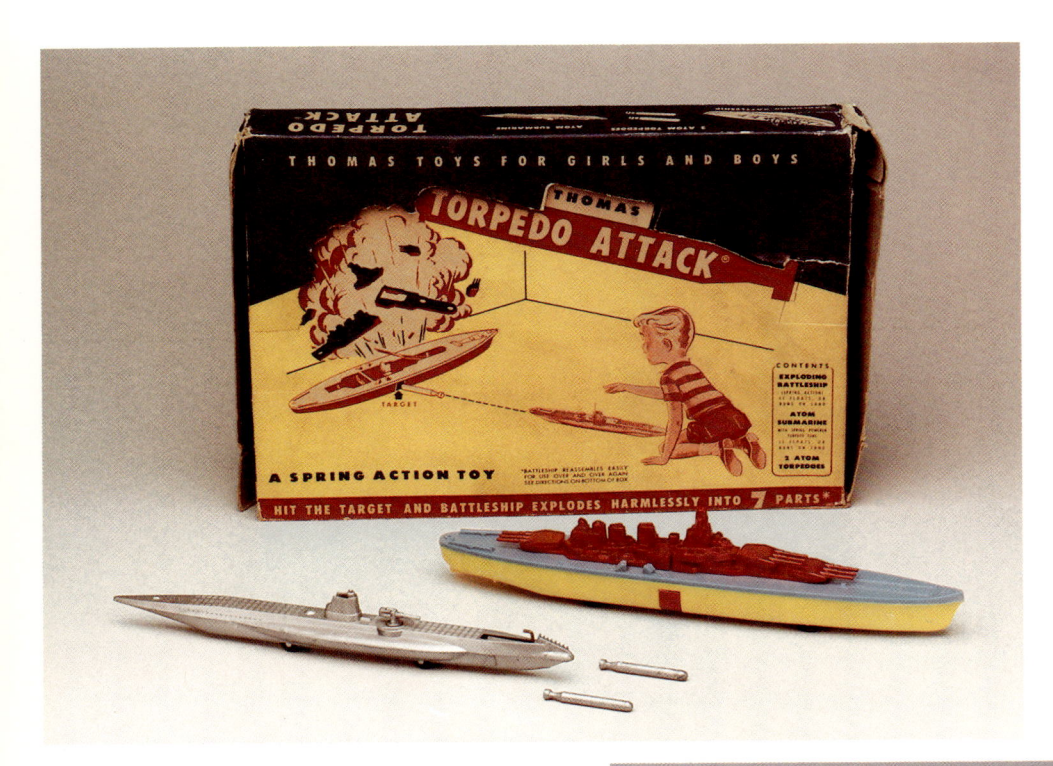

Torpedo Attack battleship, 1-3" x 2-1/2" x 1-1/2"; submarine, 11-3/4" x 1-1/2" x 4-1/2", torpedo 2-1/8"; when torpedo is fired and hits target on side of battleship it explodes into seven parts; two torpedoes included, assorted color combinations, rubber wheels, Thomas Manufacturing Corp., USA (No. 241), 1953-1955. Suggested Retail $1.98.

Torpedo Boat, 8" x 2-3/4" x 1-3/4", assorted color combinations, Thomas Manufacturing Corp., USA (No. 305), 1953 to late 1950s. Suggested Retail $0.29.

Sailboat Dinghy, 8-1/2" x 2-3/4" x 10", removable vinyl sail with wooden mast and polyethylene rudder, sold with and without swimmer, assorted colors, Thomas Manufacturing Corp., USA (No. 143), 1953 to late 1950s. Suggested Retail $0.29.

Sailboat, 4-1/2" x 1-1/2" x 4-1/4", with polyethylene sails, assorted colors, Thomas Manufacturing Corp., USA (No. 269), 1953 to late 1950s. Suggested Retail $0.10.

Cabin Cruiser, 8-1/4" x 2-3/4" x 2-1/4", assorted color combinations, Thomas Manufacturing Corp., USA (No. 350), 1953 to late 1950s. Suggested Retail $0.29.

Boat Yard Set, display box 15-3/8" x 10-1/2" x 3-3/8", assorted ships in assorted color combinations, Thomas Manufacturing Corp., USA (No. 253), 1954 to late 1950s. Suggested Retail $1.98.

BOATS 167

Sailboat with Cabin, 14" x 18-3/4" x 4-1/4", removable wooden mast with vinyl sail and polyethylene rudder, assorted color combinations, Thomas Manufacturing Corp., USA (No. 282), 1954-1956. Suggested Retail $0.98.

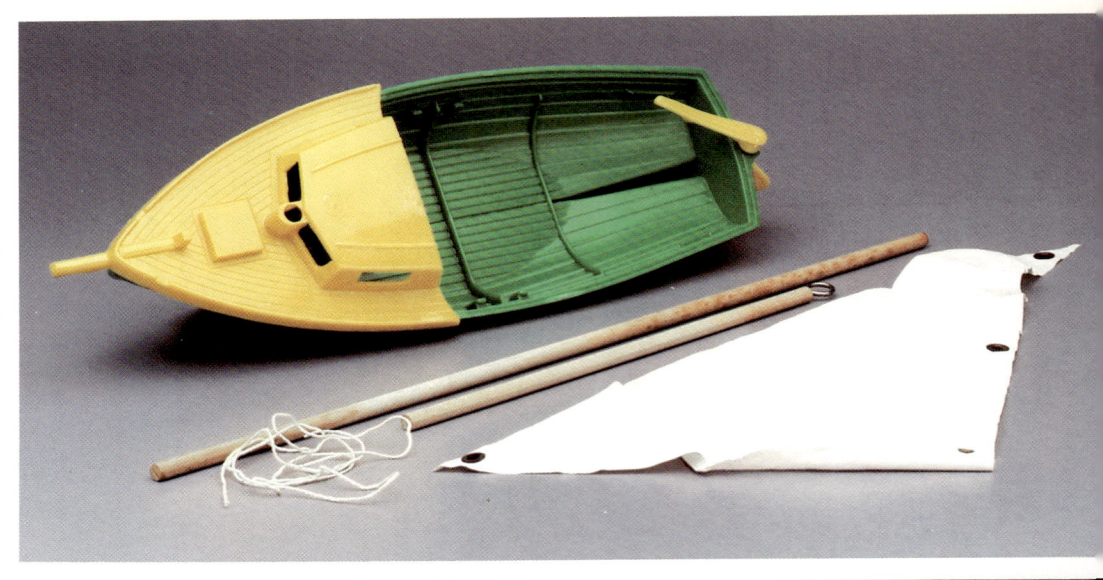

Bathtub Fleet, display box 10" x 10" x 2-1/8", assorted ships in assorted color combinations, Thomas Manufacturing Corp., USA (No. 344), 1956 to late 1950s. Suggested Retail $0.98.

Tugboat, 8-1/4" x 2-3/4" x 3", assorted color combinations, Thomas Manufacturing Corp., USA (No. 349), 1954-1956. Suggested Retail $0.29.

Weekend Cruise, card size 13" x 7" x 7-1/2", Car 4-1/4" long, Boat 4-1/2" long, Thomas Mfg. Corp., USA (No. 339), mid 1950s. Suggested Retail $0.39.

168 PLASTIC TOYS

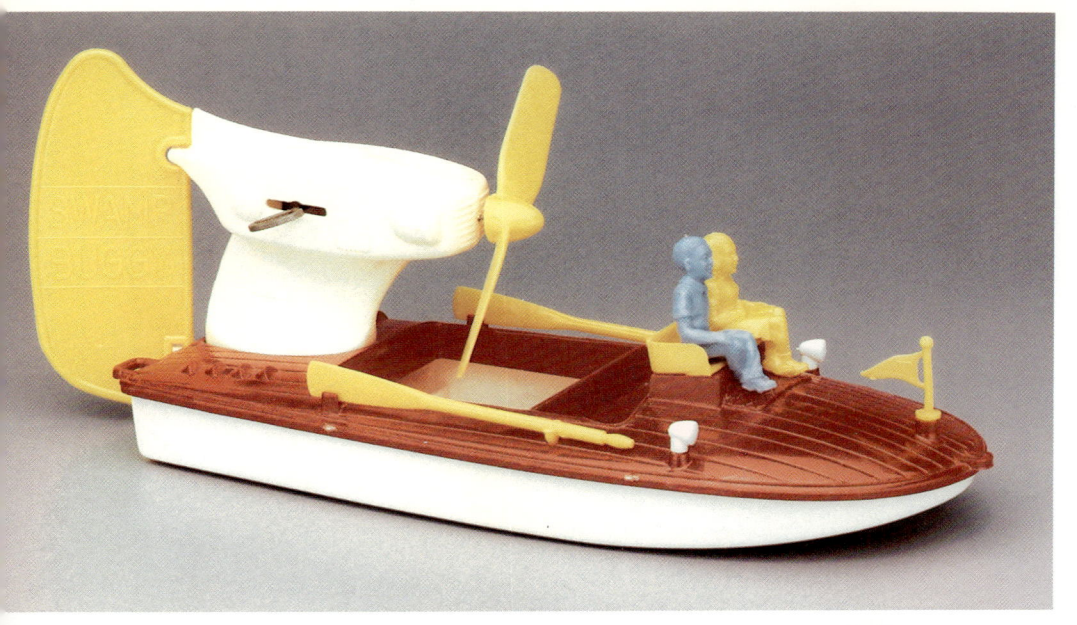

Motorized Swamp Buggy, 12-1/2" x 4" x 5", windup motor with attached key, adjustable rudder, boy, girl and dog, removable oars, assorted color combinations, Thomas Manufacturing Corp., USA (No. 487), late 1950s.

Speedboat Jeep and Trailer, racing hydroplane 8-1/2" x 3-1/4" x 3-1/2", rubber band powered, trailer 8-1/2" x 3-1/4" x 1-1/2", jeep 8" x 3-1/2" x 4", with folding windshield, assorted colors and color combinations, plastic wheels, jeep and trailer are polyethylene, Thomas Manufacturing Corp., USA (No. 545), late 1950s. Suggested Retail $1.59. Hydroplane alone, (No. 629), late 1950s.

PT Boat, 8" x 3" x 3", wind-up motor with attached key, two revolving polyethylene guns, assorted color combinations, Thomas Manufacturing Corp., USA, late 1950s (example shown missing rudder).

Floating Submarine Kazoo, 6-1/2" x 1-3/8" x 1-5/8", 20th Century Products, USA, early 1950s.

BOATS 169

Ocean Liner, 6-5/8" x 1-1/2" x 1-1/4", assorted color combinations, Manufacturer Unknown, USA, early 1950s.

Ocean Liner, 6" x 2" x 2", assorted colors, Manufacturer Unknown, USA, early 1950s. Suggested Retail $0.10.

Speedboat, 8" x 3-1/4" x 2-1/4", assorted color combinations with painted details, Manufacturer Unknown, USA, early 1950s.

Canoe, 6" x 1-3/4" x 1", assorted color combinations, Manufacturer unknown, USA, early 1950s.

CHAPTER 11
AIRPLANES AND HELICOPTERS

When the Wright brothers made their historic flight on December 17th 1903, a whole new toy category was created. Toy airplanes made from tin, cast iron and lead became popular playthings during the 1920's and 1930's.

The earliest documentation of an injection molded plastic toy airplane appears in the October, 1938 issue of *Modern Plastics Magazine*. An ad promoting Lumarith, a cellulose acetate from the Celluloid Corporation shows a DC-3 injection molded by the Kilgore Mfg. Co. of Westerville, Ohio. The DC-3 was part of Kilgore's "Jewels For Playthings" line of plastic toys.

It was molded in one piece with very thin wall sections that made it highly susceptible to breaking and warping. It also had rubber propellers and wheels that became brittle and broke, further compounding the problem of poor design.

The DC-3 is not believed to have been produced after the war and pre-war examples are very rare today.

The Dillon Beck Mfg. Co. of Hillside, New Jersey introduced the next injection molded plastic toy airplane in 1941 under their new "Wannatoy" label.

Based on another popular plane of the times, their model of the Bell P-39 Airacobra was an instant hit. With its cellulose acetate propeller, wheels and two piece fuselage, it was far superior than its earlier rival and Dillon Beck was soon flooded with orders. Unfortunately, a dwindling supply of plastic for non-military uses convinced Dillon Beck to turn to 100% defense work by the end of 1942.

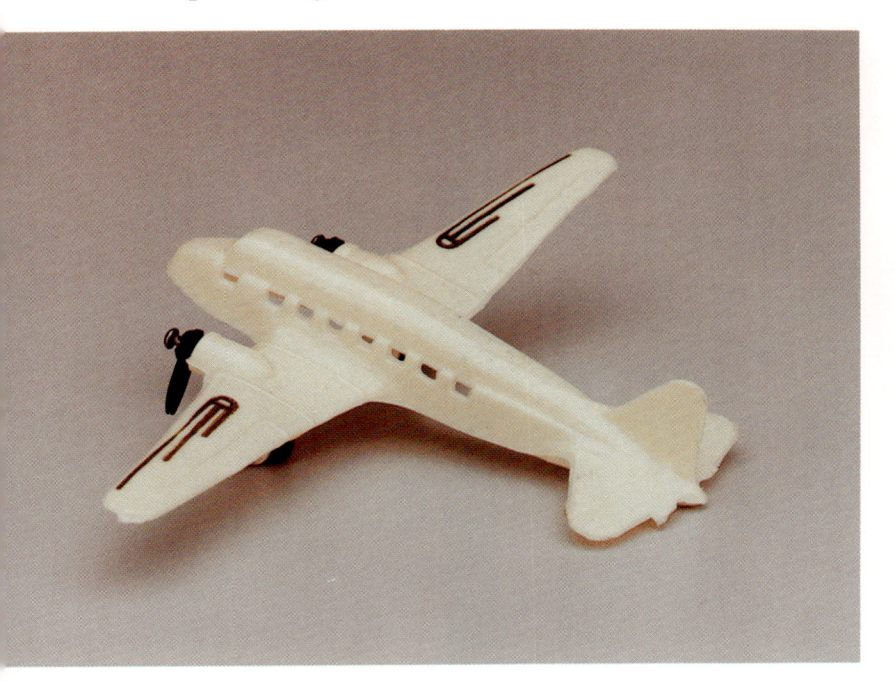

DC-3, 3-1/2" x 5" x 1", revolving rubber propellers, assorted colors with gold hot stamping, rubber wheels, Kilgore Mfg. Co., USA, 1938 to early 1940s.

P-39 Airacobra, 4-3/8" x 5" x 1-1/2", revolving propeller, red, Dillon Beck Mfg. Co., USA, (No. T20), 1941 to 1942.

The success of the P-39, however brief, caught the attention of Benjamin Shapiro, owner of Acme Plastics Manufacturing Co. of New York who contracted Consolidated Molded Products Corp. of Scranton, Pennsylvania to begin work on the molds for three airplanes and three boats. Under the watchful eye of Islyn Thomas, who was Chief Engineer at Consolidated, molds were made for a Douglas 20 Bomber, a P-40 Warhawk and a Boulton Paul Defiant.

Shapiro had intended for Consolidated to also manufacture the airplanes and boats but they converted to defense work by the end of 1941, before any could be molded. As luck would have it, Shapiro's friend, Thomas, left Consolidated for a position as General Manager at Ideal shortly after the molds were completed. In his new position, Thomas was able to put together a deal that would benefit both Acme and Ideal.

1948 Thomas Mfg. Corp. catalog.

Bell P-39 Airacobra Comparison, original 1941 Dillon Beck version on the left and the 1944 Ideal conversion on the right.

P-40 Box, 4" x 4" x 1 1/2", Acme Plastics Mfg. Co., USA, 1945-1950.

Douglas A-20 Bomber, Boulton Paul Defiant and P-40 Warhawk, Acme Plastics Mfg. Co., USA, 1945 to mid 1950s.

172 PLASTIC TOYS

Ideal agreed to mold the three boats for Acme as long as their supply of plastic permitted. In exchange, Acme would allow Ideal to temporarily market the boats in sets, thus providing Ideal's new plastics division with its first toys for boys. In another deal, the quick thinking Thomas was able to purchase the mold for the Dillon Beck P-39 and add it to the Ideal line for 1944.

Thomas remembers that Ideal had an excellent engraver on staff at the time who added considerable detail and the Ideal logo to the cavities of the Dillon Beck mold. For collectors lucky enough to have both versions of the P-39, it is easy to see that both are otherwise one and the same. The Ideal P-39 was produced from 1944 until 1947 and can still be found today. The Dillon Beck version, produced for less than two years, is quite rare.

By the end of 1944, Thomas had decided to leave Ideal and start his own company, the Thomas Manufacturing Corporation. Shapiro, who had worked with Thomas since the mid 1930s, became an equal partner and contributed both cash and the molds for the three boats and three airplanes to the new venture. These boats and airplanes, along with a jeep and helicopter designed by Thomas, made up the original Thomas Toy line, first offered in 1945.

Thomas' working model of a Bell helicopter was the first injection molded plastic toy helicopter made. Granted U.S. patent No. 2,411,596 in 1946, this best selling toy featured a propeller that revolved as it was pulled along.

Bell Helicopter, Thomas Mfg. Corp., USA (No. 12), 1945-1951.

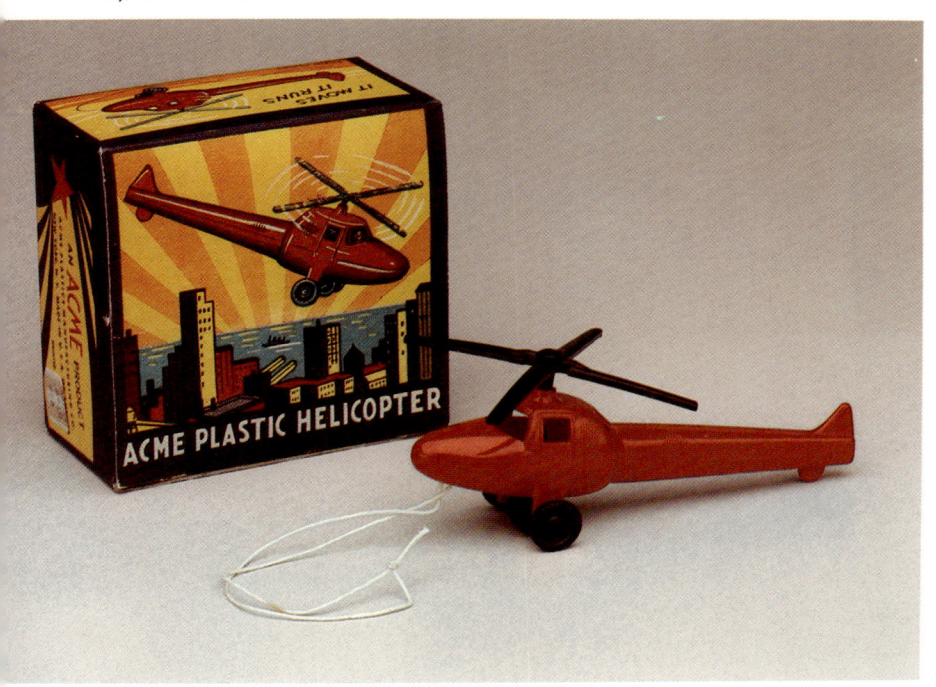

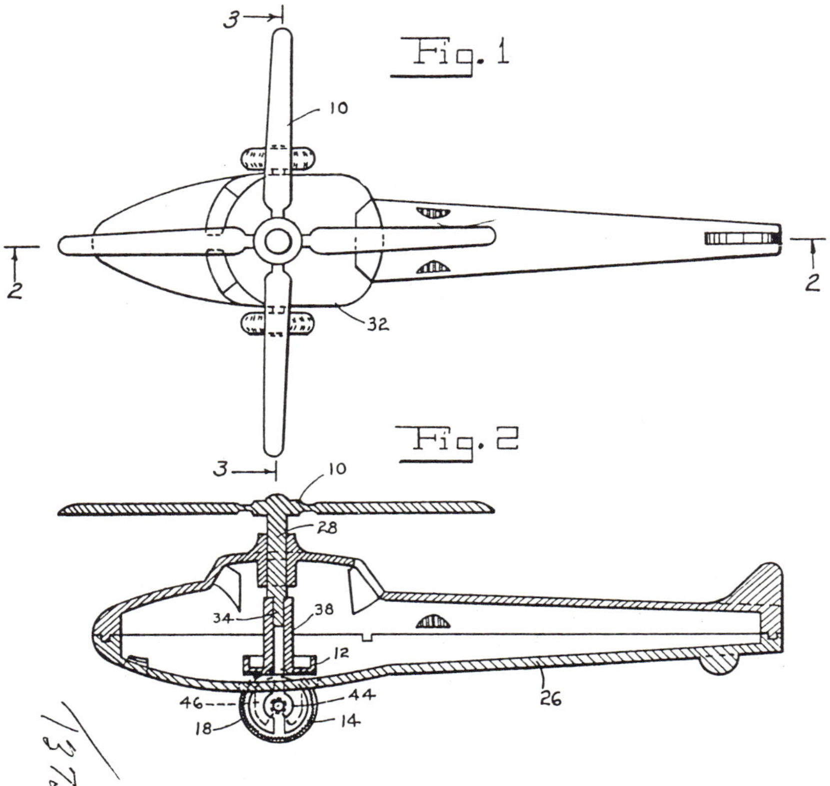

Bell Helicopter patent drawings, Thomas Mfg. Co., USA, 1946.

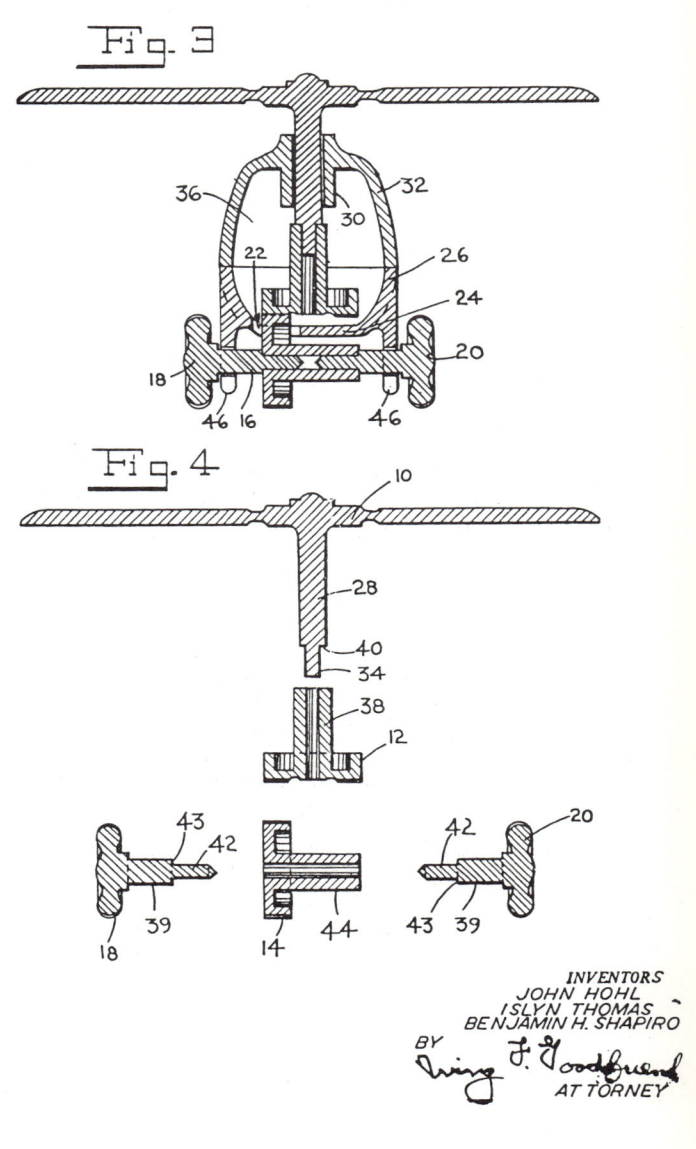

With Thomas gone, Ideal turned to its counterpart in the U.K., The Reliable Plastics Co. of Toronto, Ontario, Canada for help in those early postwar years. The histories of these two companies paralleled each other closely. Ideal was the largest manufacturer of dolls and plush in the U.S., and Reliable was the largest in the U.K. Both started injection molding toys in the early 1940's and the principals of each were close personal friends with a long history of cooperation.

When the upstart newcomer, Renwal, offered an amazing assortment of seven different "True-Scaled Planes" and five rooms of "Jolly Twins Doll House Furniture" in 1945, Ideal was able to hold its own by offering eight different airplanes, with help from Reliable.

With the exception of the P-39 and the P-40, it appears the rest of Ideal's airplane line for 1945 came from molds made by Reliable. Examples of some of the planes have been found with both companies' names on them, and some even say "made in USA" and "made in Canada" on the same piece!

By 1952, toy jets had become the rage and replaced all but a few of the propeller driven models. Those that remained somehow managed to hang on well into the mid 1950s. While their automotive counterparts delighted potential buyers with one innovation after another, plastic toy planes evolved without much fanfare.

In 1947, the Thomas Mfg. Corp. tried to make jets by removing the propellers and adding wing tanks to some of its P-40 and Defiant molds. These mutants were not accepted by the jet conscious youth of that period and were dropped the following year.

In 1950, the Hubley Mfg. Co. introduced the first plastic toy plane with retractable landing gear.

In 1951, the Ideal Toy Corp. introduced the first all new plastic toy jets and the first plastic toy bomber that could drop two plastic bombs when a button was pushed. That same year, Ideal also introduced their version of a Lockheed Model 49 Constellation with a removable transparent fuselage that allowed the owner to see the interior of the plane including the pilot, co-pilot and two stewardesses.

In 1953, Louis Marx and Co., Inc. went Ideal one better and introduced a Douglas DC-4 with a removable transparent fuselage and a crew and passengers that could be moved around inside.

In 1953, the Thomas Mfg. Corp. introduced "Bomb-A-Ship" which included an "Atom Bomber" that could drop

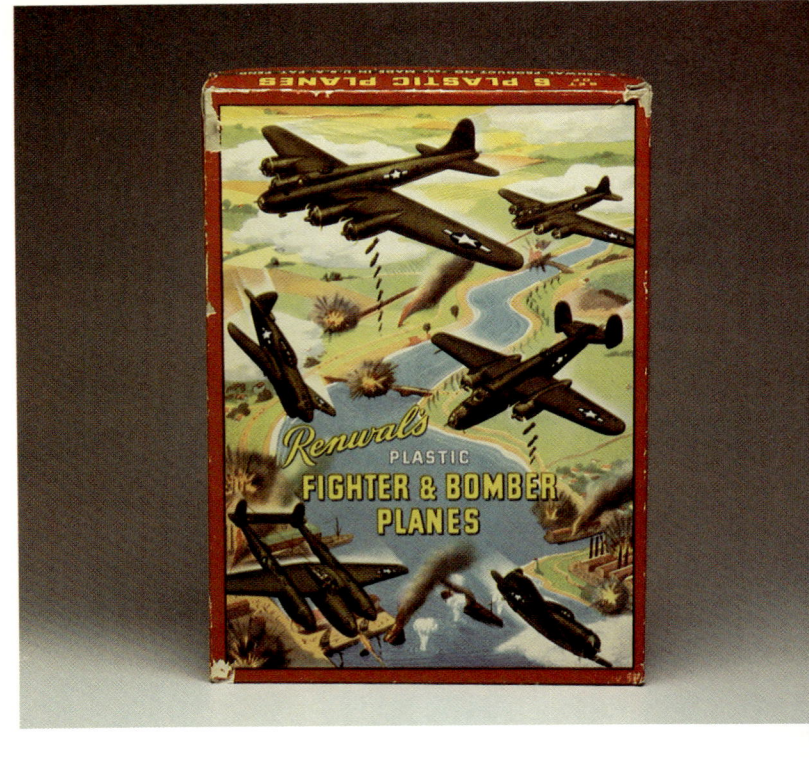

Fighter and Bomber Planes, box 17 1/4" x 12 1/2" x 2 1/2", includes; Flying Fortress (No. 777), B-17 Flying Fortress (No. 17), P-40 Curtis Warhawk (No. 40), B-25 Mitchell (No. 25), P-38 Lightning (No. 38), P-47 Thunderbolt (No. 47), Renwal Manufacturing Co., Inc., USA (No. 666), 1945-1947.

a metal bomb onto a battleship which exploded into seven pieces.

In 1957, the Ideal Toy Corp. introduced a "Flying Box Car" with a wingspan of 22 inches that contained a twenty man combat team, a cannon and five vehicles, making it the most ambitious offering of the decade.

A general lack of innovation as far as domestic plastic toy planes were concerned kept most retail prices between ten and twenty-nine cents throughout the 1950s. Because import duties were based on the wholesale value of a toy, Japan and Hong Kong concentrated their efforts on inexpensive toys such as these. The plastic toy plane was easy prey for foreign competition and by the 1960s another American innovation would all but fade away.

Shooting Jet, 5-1/2" x 5-3/8" x 1-3/8", when jet is pushed along on the floor, it fires BBs (small metal balls), assorted color combinations, rubber wheels, Argo, USA, 1953 to mid 1950s.

Jet, 5" x 4" x 1-3/8", assorted colors including olive drab, plastic wheels, Dillon Beck Mfg. Co., USA, 1951 to mid 1950s.

P-39 Airacobra, 4-1/4" x 5" x 1-1/2", revolving propeller, assorted colors with or without wing decals, plastic wheels, Ideal Novelty and Toy Co., USA (No. P39), 1944-1947. Suggested Retail $0.25.

P-40 Warhawk, 4" x 4-1/2" x 1-3/8", revolving propeller, assorted colors, plastic wheels, Ideal Novelty and Toy Co., USA (No. P40), 1945-1947. Suggested Retail $0.15.

P-51 Mustang, 4" x 4-3/4" x 1-3/8", revolving propeller, assorted colors, plastic wheels, Ideal Novelty and Toy Co., USA (No. P51), 1945-1947.

AIRPLANES AND HELICOPTERS

F4F Wildcat, 4" x 4-3/4" x 1-3/8", revolving propeller, assorted colors, plastic wheels, Ideal Novelty and Toy Co., USA (No. F4F), 1945-1947.

11D Hurricane, 4" x 4-3/4" x 1-3/8", revolving propeller, assorted colors, plastic wheels, Ideal Novelty and Toy Co., USA (No. 11D), 1945-1947.

P-47 Thunderbolt, 4" x 4-1/2" x 1-3/8", revolving propeller, assorted colors, plastic wheels, Ideal Novelty and Toy Co., USA (No. P47), 1945-1947.

DC-4 Four Motor Transport, 5-3/4" x 8" x 1-1/2", revolving propellers, assorted colors, plastic wheels, Ideal Novelty and Toy Co., USA (No. DC4), 1945-1947.

B-29 Superfortress, 5-3/4" x 8" x 1-5/8", revolving propellers, assorted colors with water decals, plastic wheels, Ideal Novelty and Toy Co., USA (No. B29), 1945-1947.

Helicopter with Single Rotor, 5" x 2-1/2" x 1-1/8", when pushed or pulled rotor rotates, assorted color combinations, plastic wheels, Ideal Novelty and Toy Co., USA (No. HC-1), 1947-1948.

Helicopter with Twin Rotors, 6" x 3-1/4" x 1-3/8", when pulled, tail propeller and twin rotors turn in opposite directions, assorted color combinations, plastic wheels, Ideal Novelty and Toy Co., USA (No. HC-3), 1947-1948.

Buzzy Airplane Pull Toy, 7" x 6 1/2" x 3 1/2", when pulled along chick bobs head up and down, assorted color combinations with painted details, plastic wheels, Ideal Toy Corporation, USA (No. 4212), 1951-1952. Suggested retail $1.00.

F-80 Panther Jet Fighter Plane, 6-1/2" long, assorted colors, plastic wheels, Ideal Toy Corporation, USA (No. 4858), 1951-1952, in olive drab assortment with F-90 (No. 4929), 1951. Suggested Retail $0.15.

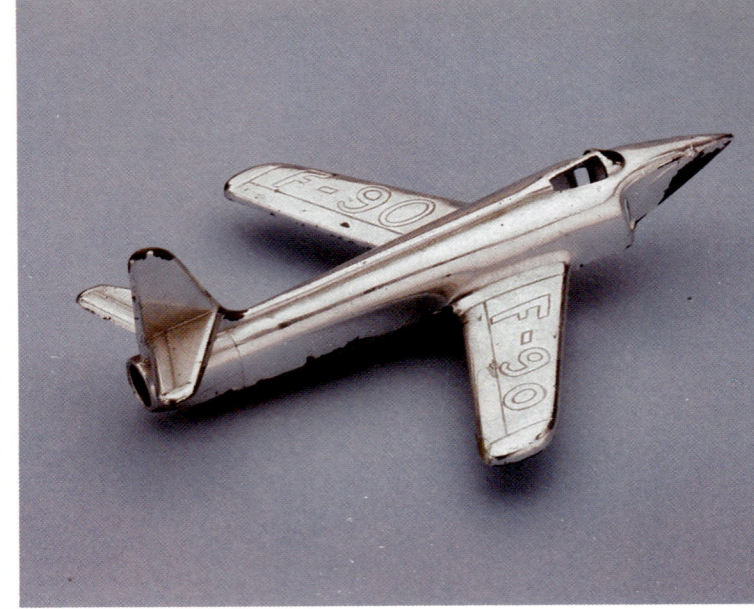

F-90 Shooting Star Jet Fighter Plane, 6-1/2" long, assorted colors, plastic wheels, Ideal Toy Corporation, USA (No. 4859), 1951-1952, in olive drab assortment with F-80 (No. 4929), 1951. Suggested Retail $0.15.

B-25 Army Bomber, 7-3/8" x 8-3/4" x 2-5/8", when dome is pressed bomb bay doors open releasing two bombs, assorted colors, plastic wheels, Ideal Toy Corporation, USA (No. 4931), 1951-1954. Suggested Retail $0.80.

Constellation Plane with Removable Transparent Fuselage, 9-1/2" x 11" x 3-1/4", with pilot, co-pilot and two stewardesses, assorted colors, rubber front wheels, plastic rear wheel, Ideal Toy Corporation, USA (No. 4841), 1951-1954. Suggested Retail $1.00.

Opposite page:
Flying Box Car and Combat Team, 18" x 22" x 5-1/2", nose of cargo plane opens and ramp is included for loading and unloading combat team, includes the following vinyl equipment; radar trailer with rotating scanner, howitzer with elevating mechanism and shells, Nike missile, trailer with launcher and missiles (not shown), searchlight trailer with rotating search-

Tornado Jet Bomber 4-1/4" x 4-15/16", assorted colors including "Silvertone", plastic wheels, Ideal Toy Corporation, USA, part of Jet Plane assortment, 12 dozen per carton (No. 4851), 1951-1954. Suggested Retail $0.05.

F-86 Sabre Jet Fighter, 4-1/4" x 4", assorted colors including "Silvertone", plastic wheels, Ideal Toy Corporation, USA, part of Jet Plane Assortment (No. 4851), 1951-1954. Suggested Retail $0.05. Not shown, Scorpion Jet Fighter, 4-1/4" x 5".

Mechanical Sea Plane, 9-1/2" long fuselage, 10" wing span, wind-up motor with attached key glides plane along the water, unbreakable polyethylene propeller, yellow and blue, with N 48475 on wing, Ideal Toy Corporation, USA, (No. 4847), 1954-1956. Suggested Retail $1.50. (No. 4847-0), 1959-1966 and 1968-1975, no number on wing 1976-1977. Reissued as Naval Rescue Plane, white and blue with U.S. Navy on wing, (No. 3151-8) 1964-1967. Reissued as U.S. Marine Air-Sea Rescue Plane, green and white with USMC on wing, (No. 3151-8), 1966-1977. Reissued as Military Seaplane in olive drab (No. 4848), possibly only one year, late 1950s.

light, truck cab and jeep to tow equipment and vinyl soldiers. Plastic fuselage, polyethylene propellers, cowlings and tailsections, lithographed tin wings, rubber wheels, blue fuselage with silver lithographed wings, olive drab and blue accessories, Ideal Toy Corporation, USA, (No. 4869), 1957-1959. Suggested Retail $6.00.

AIRPLANES AND HELICOPTERS 179

U.S. Army Plane, 5" x 6" x 2", revolving propeller and retractable landing gear, assorted color combinations, rubber wheels, Hubley Mfg. Co., USA, 1950 to late 1950s. Suggested Retail $0.29.

Twin Engine Bomber, 6" x 8" x 1-1/2", revolving propellers and retractable landing gear, assorted color combinations, rubber wheels, Hubley Mfg. Co., USA, early to mid 1950s.

Assorted Small Airplanes, 1-1/2" to 3" wingspan, assorted colors, stationary wheels, Lido Toy Corp., USA, mid 1940s to late 1950s.

Four Airforce Airplanes, set includes two jets, 6-1/4" x 4-1/4" x 2-1/4" and 5" x 6" x 1-1/2", two propeller driven planes, 5" x 5-3/4" x 1-1/2" and 5" x 5-3/4" x 1-3/8" with revolving propellers, plus two hauling tractors, 2-5/8" x 1-1/2" x 1-3/4", olive drab; planes have stationary wheels, tractors have plastic wheels that turn, Louis Marx and Co., Inc., USA, 1951 to mid 1950s.

Seven Different Airplanes, set includes one jet, 3-3/4" x 3-7/8" x 1-1/4" and six propeller driven planes, average size 3-3/4" x 4-1/4" x 1", with stationary propellers, plus two hauling tractors, 2" x 3/4" x 1", olive drab; planes and tractors have stationary wheels, Louis Marx and Co., Inc., USA, 1951 to mid 1950s.

Assorted Small Planes, set includes eleven different planes, nine propeller driven and three jets, 3" to 4" wingspans, olive drab, stationary wheels, Louis Marx and Co., Inc., USA, 1951 to mid 1950s.

B-47 Stratojet, 5-1/2" x 5-1/2" x 1-1/2", assorted colors, plastic wheels, Louis Marx and Co., Inc., USA, early 1950s.

AIRPLANES AND HELICOPTERS 181

FIX-ALL DC-6 Passenger Plane, 11-1/2" x 17" x 4", take apart plane with over thirty pieces, removable clear top half of fuselage, pilot, copilot, two hostesses and light passengers that may be moved about in the cabin, assorted color combinations with painted wing markings and decal on tail, plastic wheels, Louis Marx and Co., Inc., USA, 1953 to mid 1950s. Suggested Retail $1.98.

F2H2 Banshee, 7-1/2" x 8-1/4" x 1-3/4", a Marx playset jet, metallic silver, plastic wheels, Louis Marx and Co., Inc., USA, mid 1950s.

F-86 Sabre, 8" x 8-1/2" x 2-1/4", a Marx playset jet, metallic silver, plastic wheels, Louis Marx and Co., Inc., USA, mid 1950s.

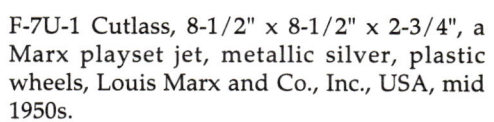

F-7U-1 Cutlass, 8-1/2" x 8-1/2" x 2-3/4", a Marx playset jet, metallic silver, plastic wheels, Louis Marx and Co., Inc., USA, mid 1950s.

Jet, 4" x 4" x 1-1/4", assorted colors including metallic silver, plastic wheels, Premier Products Co., USA, early to mid 1950s.

Jet, 3-3/4" x 4" x 1-1/8", assorted colors including metallic silver, plastic wheels, Premier Products Co., USA, early to mid 1950s.

Jet, 2-3/4" x 2-3/4" x 5/8", assorted colors including olive drab, plastic wheels, Premier Products Co., USA, early to mid 1950s.

Flying Helicopter, 6-3/4" x 3-3/4" x 2-1/2", when helicopter is pushed up a 12" long spiral rod, aluminum blades spin and helicopter takes off flying, as high as seventy five feet in the air, assorted color combinations, Pre-View Toy and Novelty Co., Inc., USA (No. 500), 1946 to late 1940s. Suggested Retail $0.75.

AIRPLANES AND HELICOPTERS

Jet, 7-3/4" x 4-1/4" x 2-1/4", olive drab, metallic silver and assorted color combinations with painted wing markings, plastic wheels, Pyro Plastics Corp., USA, early to mid 1950s.

F7U-Cutlass, 6" x 6" x 2", olive drab, metallic silver and assorted color combinations with painted wing markings, plastic wheels, Pyro Plastics Corp., USA, early to mid 1950s.

Jet, 1-1/2" x 1-3/8" x 3/8", olive drab with painted white wing markings, stationary wheels, Pyro Plastics Corp., USA, early 1950s.

B-17 Flying Fortress, 3-3/4" x 5" x 3/4", revolving propellers, assorted colors, plastic wheels, Renwal Manufacturing Co., Inc. USA (No. 17), 1945-1947.

184 PLASTIC TOYS

P-40 Curtis Warhawk, 4-1/4" x 4-3/4" x 1-1/2", revolving propeller, assorted colors and color combinations, plastic wheels, Renwal Manufacturing Co., Inc. USA (No. 40), 1945-1947, 1949-1955. Suggested Retail $0.10.

P-47 Thunderbolt, 5-3/4" x 6-1/2" x 1-3/4", revolving propeller, assorted colors with decals, plastic wheels, Renwal Manufacturing Co., Inc. USA (No. 47), 1945-1947.

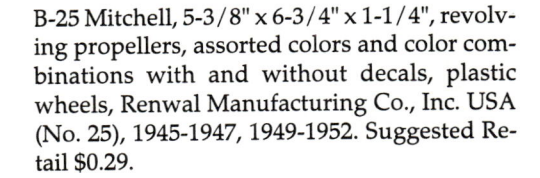

B-25 Mitchell, 5-3/8" x 6-3/4" x 1-1/4", revolving propellers, assorted colors and color combinations with and without decals, plastic wheels, Renwal Manufacturing Co., Inc. USA (No. 25), 1945-1947, 1949-1952. Suggested Retail $0.29.

AIRPLANES AND HELICOPTERS 185

Flying Fortress, 7-1/16" x 9-3/4" x 2", revolving propellers, assorted colors and color combinations with and without decals, plastic wheels, Renwal Manufacturing Co., Inc. USA (No. 777), 1945-1947, 1949-1952. Suggested Retail $0.49.

B-29 Super Fortress, 5-3/8" x 7" x 1-1/4", revolving propellers, assorted colors and color combinations with and without decals, plastic wheels, Renwal Manufacturing Co., Inc. USA (No. 29), 1945-1947, 1951-1954. Suggested Retail $0.29.

P-38 Lightning, 5-1/8" x 6-7/8" x 1-1/4", revolving propellers, assorted colors and color combinations with and without decals, plastic wheels, Renwal Manufacturing Co., Inc. USA (No. 38), 1945-1947, 1949-1953. Suggested Retail $0.29.

DC-4 Transport Plane, 7 1/8" x 5 5/8" x 1 1/8", revolving propellers, assorted color combinations with and without decals, plastic wheels, Renwal Manufacturing Co., Inc., USA (No. DC-4), 1949-1956. Suggested retail $0.29.

Martin Mars Flying Boat, 6 3/4" x 5 1/2" x 1 1/2", revolving propellers, assorted color combinations, plastic wheels, Renwal Manufacturing Co., Inc., USA (No. 15), 1949-1955. Suggested retail $0.29.

Airport Plane Set, box 15-3/4" x 10-1/4" x 3-1/2", seven piece set includes (No. 4) DC-4, (No. 15) Martin Mars Flying Boat, (No. 25) Mitchell Bomber, (No. 29) Bomber Plane, (No. 38) Lightning Fighter, (No. 40) Curtiss Warhawk and (No. 777) Plane, assorted colors and color combinations, plastic wheels, Renwal Manufacturing Co., Inc. USA (No. 366), 1951, (Nos. 25, 38, 40, and 777 replaced by Nos. 161, 162, 164, and 166), 1952-1954. Suggested Retail $1.98.

AIRPLANES AND HELICOPTERS 187

Army Thunder Jet Plane, 9-1/4" x 8-1/8" x 2", soft vinyl nose piece, assorted colors, plastic wheels, Renwal Manufacturing Co., Inc. USA (No. 161), 1952-1956. Suggested Retail $0.39. Example shown with peg on top of fuselage was made for the Advance Design Development Co. of Detroit, Michigan. Standard model does not have this peg.

Navy Panther Jet Plane, 9" x 7-3/4" x 2", soft vinyl nose piece, assorted colors, plastic wheels, Renwal Manufacturing Co., Inc. USA (No. 162), 1952-1956. Suggested Retail $0.39. Example shown with peg on top of fuselage was made for the Advance Design Development Co. of Detroit, Michigan. Standard model does not have this peg.

Army Thunder Jet Plane, 5-1/4" x 4-3/4" x 1-3/8", assorted colors, plastic wheels, Renwal Manufacturing Co., Inc. USA (No. 164), 1952-1956. Suggested Retail $0.10.

Navy Panther Jet Plane, 5-1/4" x 5" x 1-3/8", assorted colors, plastic wheels, Renwal Manufacturing Co., Inc. USA (No. 166), 1952-1956. Suggested Retail $0.10.

Fairchild XC-120, 8-1/4" x 10-1/4" x 2-1/2", revolving propellers, front and rear cargo doors open, with 1-1/2" long jeep and tank, gray with decals, rubber wheels, Swader Plastic Co., (No. XC-120) USA, 1952 through early 1950s. Suggested Retail $1.00.

Douglas A-20 Bomber, 4-1/2" x 5-1/2" x 1-1/4", revolving propellers, assorted colors, plastic wheels, Acme Plastics Mfg. Co., USA, 1942-1944, Thomas Manufacturing Corp., USA (No. 21), 1945-1953. Suggested Retail $0.10. Vacuum metalized (No. 81), olive drab (No. 186).

P40, 4" x 4-1/2" x 1-1/2", revolving propeller, assorted colors, plastic wheels, Acme Plastics Mfg. Co., USA, 1942-1944, Thomas Manufacturing Corp., USA, part of plane assortment along with Defiant (No. P40-10), 1945 to mid 1950s. Suggested Retail $0.10.

Defiant, 4" x 4-1/2" x 1-1/2", revolving propeller, assorted colors, plastic wheels, Acme Plastics Mfg. Co., USA, 1942-1944, Thomas Manufacturing Corp., USA, part of plane assortment along with P40 (No. P40-10), 1945 to mid 1950s. Suggested Retail $0.10.

AIRPLANES AND HELICOPTERS

Bell Helicopter, 4-3/4" x 3" x 1-7/8", when pulled along, propeller revolves, assorted color combinations, plastic wheels, Thomas Manufacturing Corp., USA (No. 12), 1945-1951.

Sikorsky Helicopter, 6-3/4" x 1-3/4" x 2", when pulled along, propeller revolves, assorted colors including olive drab, plastic wheels, Thomas Manufacturing Corp., USA (No. 21), 1950 to mid 1950s, (No. 187) in olive drab, 1952. Suggested Retail $0.29.

American Airlines Passenger Plane, 4-1/2" x 6-1/2" x 1-1/4", propellers turn, assorted colors including metallic, plastic wheels, Thomas Manufacturing Corp., USA (No. 94) and (No. 97) if vacuum metalized, 1950 to mid 1950s. Suggested Retail $0.15.

190 PLASTIC TOYS

Atom Bomber with Target and Bomb, Boeing Stratocruiser, 7-1/4" x 9-1/4" x 2-1/4", with opening bomb bay doors and 1-3/8" long metal bomb, revolving propellers, silver with decals, plastic wheels, Thomas Mfg. Corp., USA (No. 224), 1952-1954, sold separately with target box or with exploding battleship.

Kiddyland Airplane Ride, 11" wide x 10-3/4" high, plane with animal pilot 2-5/8" x 3-1/4" x 1-1/4"; when ball is moved to top of wire pole, two planes, one with a dog and the other with a bear as pilots, chase each other round and round as it returns to the base; assorted color combinations, Thomas Manufacturing Corp., USA (No. 278), 1954. Suggested Retail $0.59. Planes sold separately, (No. 215). Suggested Retail $0.10.

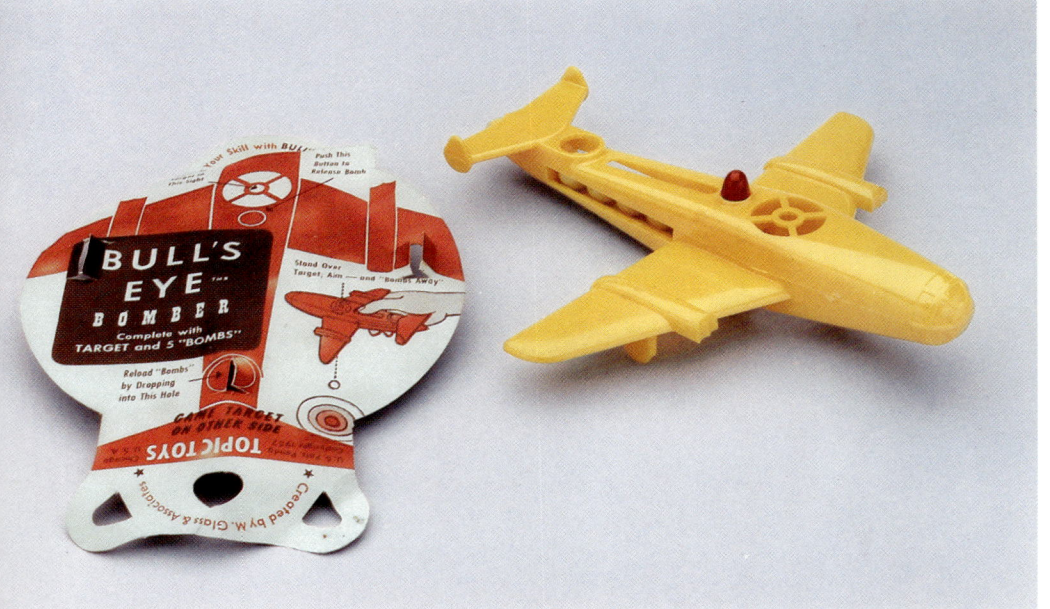

Zoom Helicopter, 4-3/4" x 8" x 2-1/4"; just wind the knob on the launcher, pull trigger and helicopter soars up to forty feet high; red and yellow, Tom and Co., USA (No. 50), 1953. Suggested Retail $0.98.

Bulls Eye Bomber, 7-3/8" x 6-3/4" x 1-5/8": when button on top is pushed, bombs drop past bomb sight, with five 1/2" diameter yellow glass marbles for bombs and packaging that forms target; assorted colors, Topic Toys, USA, 1952 to mid 1950s.

AIRPLANES AND HELICOPTERS 191

B-17, 4-3/8" x 5-1/2" x 1", revolving propellers, assorted colors, plastic wheels; B-36, 4-3/8" x 5-1/2" x 1", revolving propellers, assorted colors, plastic wheels; Jet and Globemaster, 3" to 3-3/4" wingspan, assorted colors, plastic wheels, VIBRO-ROLL Products, USA, early to mid 1950s.

C-124 Globmaster, 2-3/4" x 3-5/8" x 3/4", assorted colors, no wheels, Manufacturer Unknown, USA, early 1950s.

Constellation, 4-1/4" x 5-1/4" x 7/8", olive drab, stationary wheels, Manufacturer Unknown, USA, early 1950s.

192 PLASTIC TOYS

CHAPTER 12
SPACE TOYS

Just as another generation had been kept on the edge of their seats by the exploits of Buck Rogers and Flash Gordon, kids growing up in the 1950s would have their own inner-galactic heroes to idolize. Fueled by the media of television, comics and the movies, America once again became fascinated with outer space. Always searching for something new, toy manufacturers were quick to capitalize on the success of such television shows as "Captain Video" (1949), "Space Patrol" (1950), "Tom Corbett, Space Cadet" (1950), "Rocky Jones, Space Ranger" (1953), and "Rod Brown of the Rocket Rangers" (1953).

By 1953, toy counters across the country had taken on a new look. Helmeted space figures and rocket jet cars began to challenge the traditional best sellers in these categories, toy soldiers and race cars. Other categories were affected also. Ray guns and space helmets went head to head with six shooters and cowboy hats. Even the lowly bubble pipe took on a futuristic look with the addition of a few fins and port holes.

Information concerning the companies that manufactured these space toys has proven to be the most difficult to find of any category in this book. .With the exception of Dillon Beck, Ideal, Renwal and Thomas, it appears that none of the other companies represented here ever printed a catalog.

The Pyro Plastics Corp. of Union, New Jersey, manufacturer of what many collectors consider to be the best designed spaceships of the period, only produced a catalog for its line of hobby kits, according to President William Lester.

Without manufacturer catalogs, the historian must depend on sightings in trade journals, consumer catalogs and recollections of former employees, to date a given toy.

This method of identification brings us to Archer Plastics of New York. Archer is best known for its classic spacemen which have become the standard by which all other spacemen are judged. Most collectors agree that Archer made the first plastic spacemen. The date that these figures were introduced, however, has been the subject of hot debate.

The earliest consumer catalog sighting is from 1952. This seems to go along with the fact that Archer's futuristic vehicles are copyrighted 1952 on the bottom. The spacemen could predate the vehicles by a year or two, but until someone can produce concrete evidence the 1952 date will have to stand.

While space toys peaked as a hot category between 1952 and 1953, there is evidence that the first plastic spaceships or rocket cars appeared in 1948. An ad in *Toy News for the Toy Buyer*, a Monsanto Chemical Company publication from 1948, advertises the "Thunderbolt Rocket Car" by California Moulders Inc. of Los Angeles. Another ad, in the April, 1948 issue of *Chain Store Age*, advertises

Space People, box 13-1/2" x 10-3/4" x 1-1/2", eleven piece set includes five different 3-3/4" tall spacemen with removable helmets, three different 3-3/4" tall spacewomen and three of the same 3" tall spaceboys, assorted metallic colors, Archer Plastics, USA (No. 198), 1952 to mid 1950s. Suggested Retail $0.98.

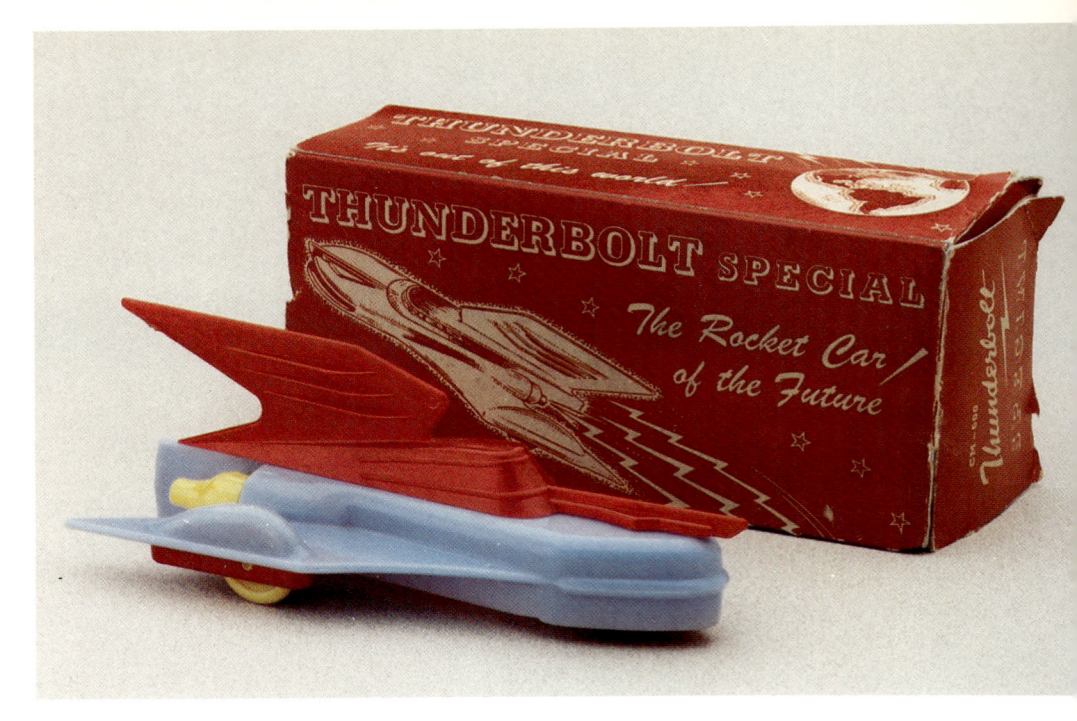

Thunderbolt Special, 7-1/2" x 3-1/2" x 2-3/4", combination of red, yellow and blue when boxed or white and pink when bagged, both with plastic wheels, California Moulders Inc., USA (No. 600), 1948 to early 1950s. Renamed Space Car in early 1950s (No. 160).

the "Aerocar" by the Plas-Tex Corp. also of Los Angeles.

The Plas-Tex Corp. produced some of the best designed plastic toys of the late 1940s and today their "Aerocar," "Hollywood Bus" and jeep are very popular among plastic collectors.

The fact that such a small toy line was able to get nationwide distribution owes a lot to the marketing genius of Frank Berlin. Berlin, who was born in Kankakee, Illinois, certainly took the roundabout way to fame and fortune in the plastics industry. In 1926, he started working for the Walgreen Drug Company in New York. By 1937 he was the Walgreen Agency's sales manager for the whole southern United States. In 1938, with eleven years of experience behind him, he decided to leave the Walgreen Agency and open up a chain of four drug stores in Florida.

In 1946, he sold the chain known as Bay Drugs to Rexall and as part of the deal, relocated in Los Angeles where he served as a Rexall vice president.

It wasn't long, however, until Berlin got the bug to open another chain of drug stores in Florida. In 1947 he left Rexall, signed leases for four new self-service drug stores, and was ready to move the family back to Florida. There was only one problem with this plan. Berlin's family didn't want to move so Berlin decided to operate the stores from California.

About this time a business associate, Scott Appleby, asked Berlin if he would like to try and turn around one of his many business ventures, the Plas-Tex Corporation. Berlin agreed, but didn't want a salary or set hours.

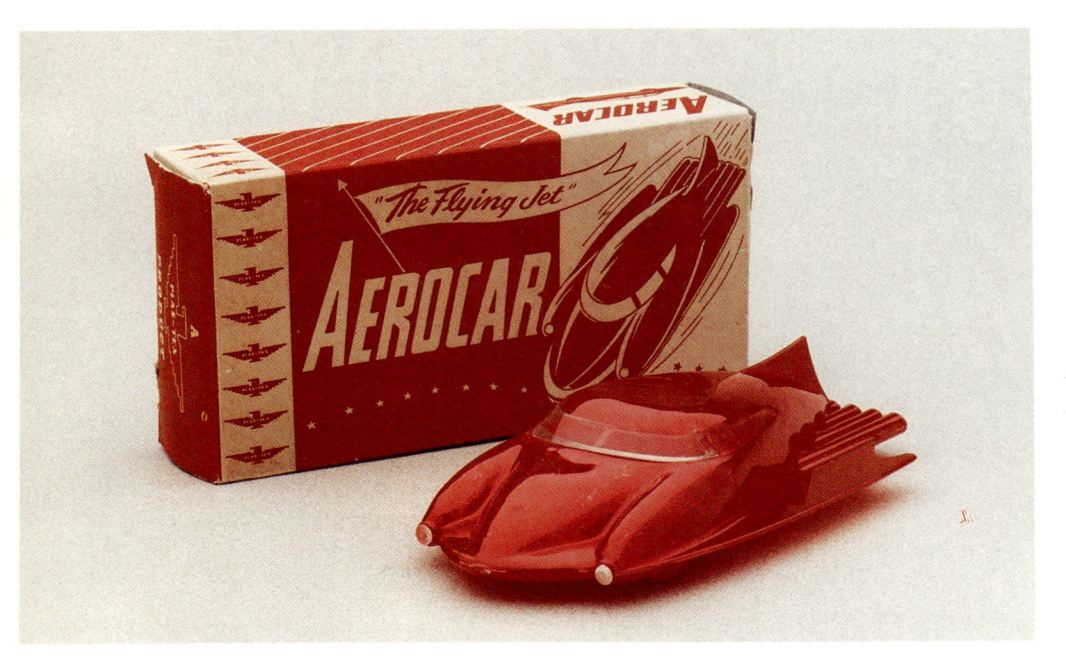

Aerocar, 7-1/2" x 4" x 1-1/2", red with plastic wheels, Plas-Tex Corp., USA (No. 56), 1948 to early 1950s.

194 PLASTIC TOYS

Plas-Tex sales had dropped from six million dollars a year during the war to fifty thousand dollars a month by the beginning of 1948 when Berlin took over as president. With twenty-six years of management and marketing experience, Berlin was on a first name basis with most of the buyers from all the major chains and used this to his advantage. By 1950, Plas-Tex's custom molding, housewares and toy sales were generating three hundred thousand dollars a month.

Seeing the potential in plastics, Berlin expressed an interest in investing in Plas-Tex. When he and Appleby could not reach an agreement, Berlin left Plas-Tex in March, 1950.

In 1951, Berlin formed a partnership with Walter McKinley, a former Plas-Tex employee who owned a struggling custom molding and tool and die shop. The name of the new enterprise was Beemak Plastics, which came from combining the last names of Berlin and McKinley. Berlin worked his magic again and had the company turned around in less than one year. The hot item was a key punch card holder for the fledgling computer industry that would eventually sell over twenty million units.

By 1952, Berlin was preparing a small line of toys for the 1953 Toy Fair. When Beemak unveiled its line that year, it included "Finicky Fido", a three inch tall dog that would flip over and land on his feet when his head was pushed into his dish, "Beemak Village", a snap together "HO scale" town similar to Bachman Brothers "Plasticville" and the "Official Space Patrol Helmet".

Weighing in at two and one half pounds, the injection molded cellulose acetate helmet came complete with "inflated double astral jet pack tanks" and cost a whopping $4.95. It was the hit of the 1953 Toy Fair. No other space toy from the period would receive as much attention as this Beemak masterpiece.

It appeared in *Colliers*, the *Saturday Evening Post* and was on the cover of Marshall Field's catalog that year. It was also written into numerous syndicated comic strips and appeared in dozens of newspaper articles throughout the country. The Santa Fe Railroad, Swifts and other companies that wanted to capitalize on the nation's fascination with outer space used it as a prop in magazine ads and as a give-away prize in numerous contests.

The ultimate compliment came when Berlin provided fifty helmets for the space men who would hold a seventy foot tall and forty foot wide helium-filled rubber "Space Man" in the 1953 Macy's Thanksgiving Day Parade. The giant "Space Man" was the hit of the parade and the Beemak helmet was the space toy of the year.

All good things must come to an end, however, and by December, 1954 America's youth were ready to turn in their space helmets for coonskin caps. The adult western would soon dominate the airways, and space toys would not be popular again until 1957, when Russia rocked the world with the launching of Sputnik.

Spacewomen and Spaceboy, women 3-3/4" tall with removable helmets, boy 3" tall; helmet will not fit woman with baby or boy, assorted metallic colors, Archer Plastics, USA, 1952 to mid 1950s, complete set is one robot, six different men, three different women and one boy.

Spacemen and Robot, 3-3/4" tall spacemen with removable helmets and 3-3/4" tall robot, two different style helmets vary slightly depending on which way the head is turned, assorted metallic colors, Archer Plastics, USA, 1952 to mid 1950s, complete set is one robot, six different men, three different women and one boy.

Spacemen, 3-3/4" tall with removable helmets, two different style helmets, vary slightly depending on which way the head is turned, assorted metallic colors, Archer Plastics, USA, 1952 to mid 1950s, complete set is one robot, six different men, three different women and one boy.

Captain Video, Video Rangers and Tobar the Robot; Captain Video and Rangers 3-5/8" tall with removable square base helmets; Tobar 4-1/8" tall, assorted metallic colors, Archer Plastics, USA, 1952 to mid 1950s. Complete set is four different poses.

Spacemen with detachable weapons and shoulder bags; six different poses, weapons and bags; weapons and bags are vinyl, 3-3/4" tall, assorted metallic colors, Archer Plastics, USA, 1952 to mid 1950s.

Space People, box 8-1/2" x 5-1/2" x 1-1/2", three piece set includes 3-3/4" tall spaceman with removable helmet, 3-3/4" tall spacewoman with baby and 3" tall spaceboy, assorted metallic colors, Archer Plastics, USA (No. 129), 1952 to mid 1950s. Suggested Retail $0.29.

Rocket Ship Kit, box 15-1/2" x 7" x 1-3/4", young space engineers can assemble and disassemble this nine piece 13" tall rocket without gluing, assorted color combinations, Archer Plastics, USA, 1952 to mid 1950s.

Cars of Tomorrow Convertible, 5" x 2" x 1-3/8", assorted colors including metallics, plastic wheels, Archer Plastics, USA, 1952 to mid 1950s.

Cars of Tomorrow Convertible, 10" x 4" x 2-5/8", with separate clear windshield, assorted colors, plastic wheels, Archer Plastics, USA, 1952 to mid 1950s.

Cars of Tomorrow Sedan, 5" x 2" x 1-1/2", assorted colors including metallics, plastic wheels, Archer Plastics, USA, 1952 to mid 1950s.

SPACE TOYS 197

Cars of Tomorrow Sedan, 10" x 4" x 3", assorted colors, plastic wheels, Archer Plastics, USA, 1952 to mid 1950s.

Cars of Tomorrow Coupe, 5" x 2" x 1-5/8", assorted colors including metallics, plastic wheels, Archer Plastics, USA, 1952 to mid 1950s.

Cars of Tomorrow Coupe, 10" x 4" x 3-1/4", assorted colors, plastic wheels, Archer Plastics, USA, 1952 to mid 1950s.

Cars of Tomorrow Gasoline Truck, 5" x 2" x 1-3/4", assorted colors including metallics, plastic wheels, Archer Plastics, USA, 1952 to mid 1950s.

198 PLASTIC TOYS

Cars of Tomorrow Gasoline Truck, 10" x 4" x 3-1/2", assorted colors, plastic wheels, Archer Plastics, USA, 1952 to mid 1950s.

Cars of Tomorrow Pickup Truck, 5" x 2" x 1-3/4", assorted colors including metallics, plastic wheels, Archer Plastics, USA, 1952 to mid 1950s.

Cars of Tomorrow Pickup Truck, 10" x 4" x 3-1/2", assorted colors, plastic wheels, Archer Plastics, USA, 1952 to mid 1950s.

Cars of Tomorrow Searchlight Truck, 10-1/2" x 4" x 5-1/2", with friction motor battery powered searchlight and 3" tall spaceboy glued to the added rear platform, assorted colors, rubber wheels, Archer Plastics, USA (No. 333), 1952 to mid 1950s.

SPACE TOYS 199

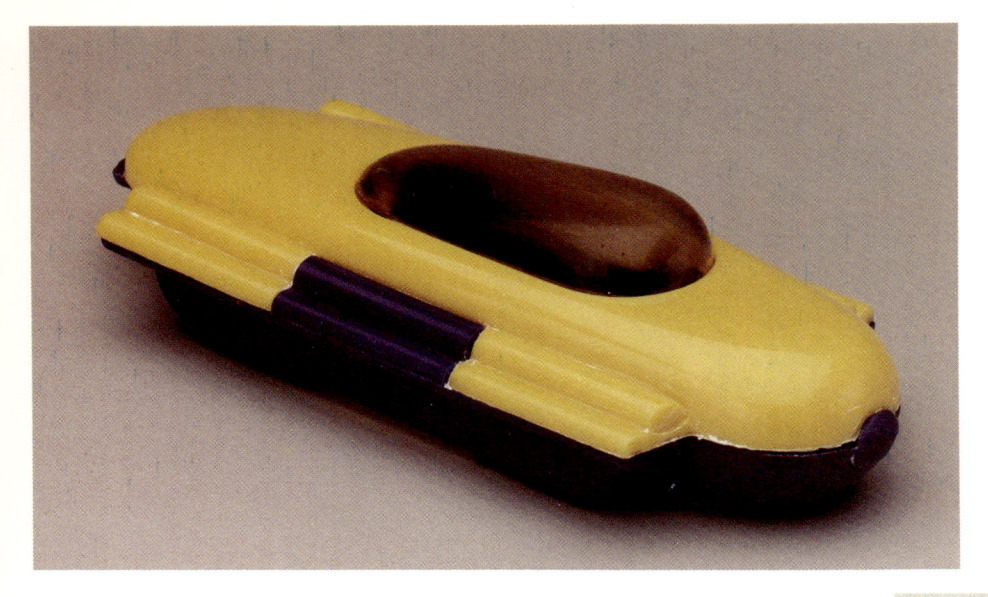

Jet Auto, 3-1/2" x 1-1/2" x 1-1/4", assorted color combinations, plastic wheels, Dillon Beck Manufacturing Company, USA (No. J-2), 1948 to mid 1950s. Suggested Retail $0.10.

Rocket Bank, 10" x 2-3/4" x 2-3/4", with removable coin trap, combination of red and yellow with or without details painted in black, sometimes found with the name of a bank when used as a giveaway, Foster Grant Co., Inc., 1952 to mid 1950s.

Rocketships and Rocket Cars, 4" long, all ten known versions shown, assorted color combinations, Gilmark Merchandise Corp., USA, early 1950s.

200 PLASTIC TOYS

Rocket Car, 4" x 2-1/4" x 1-1/2", with driver, assorted color combinations, plastic wheels, Gilmark Merchandise Corp., USA, early 1950s. Clicker Ray Gun, 5" long, red and yellow, Palmer Plastics, Inc., USA, 1953. Suggested Retail $0.10.

Futuramic Attack Transport, 11-3/4" long, cars 2-1/2" long, with driver and three removable space cars and two space machine guns that rotate, assorted color combinations, concealed plastic wheels, Ideal Toy Corporation, USA, (No. 4866), 1953-1954. Suggested Retail $0.60.

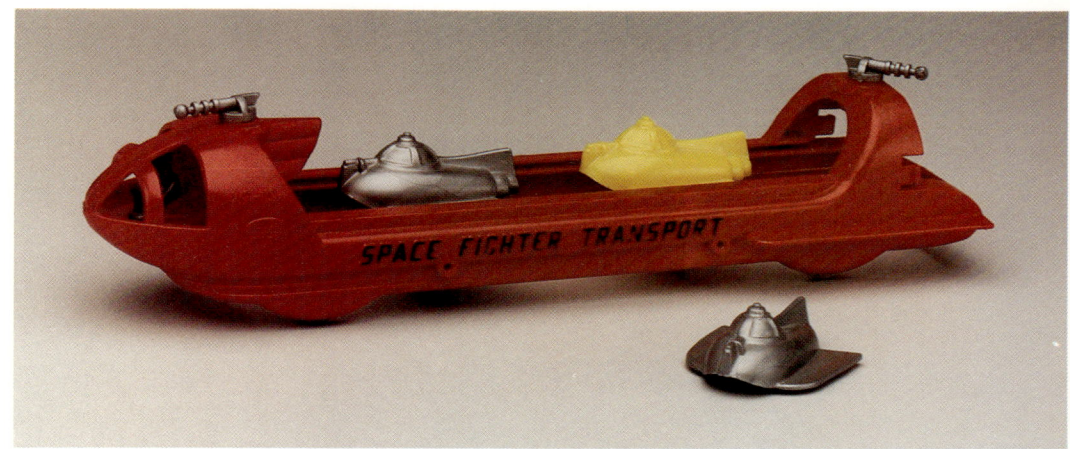

Cap Firing Sub-Machine Gun, 25-1/2" long, futuristic design with gun barrel that recoils with each rotation of hand crank, fires roll of caps, assorted color combinations, Ideal Toy Corporation, USA, (No. 4296), 1953-1954. Suggested Retail $3.00. (Removable lense on front of sight missing from example shown).

Three-Way Futurama Ray Gun, 9" long, complete with two flashlight batteries and switch for changing beam of light from red to green to white, assorted color combinations, Ideal Toy Corporation, USA, (No. 4285), 1951-1954. Suggested Retail $1.50.

Spark-Shooting Backfiring Car, 11-5/8" x 5" x 3-3/8", when trigger on end of cable is pulled, sparks shoot out of rear as roll of caps is fired, replaceable flint, assorted colors, rubber wheels, Ideal Toy Corporation, USA, (No. 4050), 1953-1954. Suggested Retail $3.00.

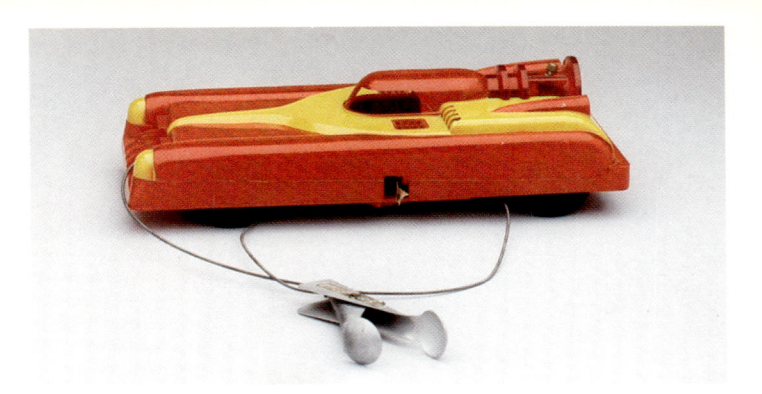

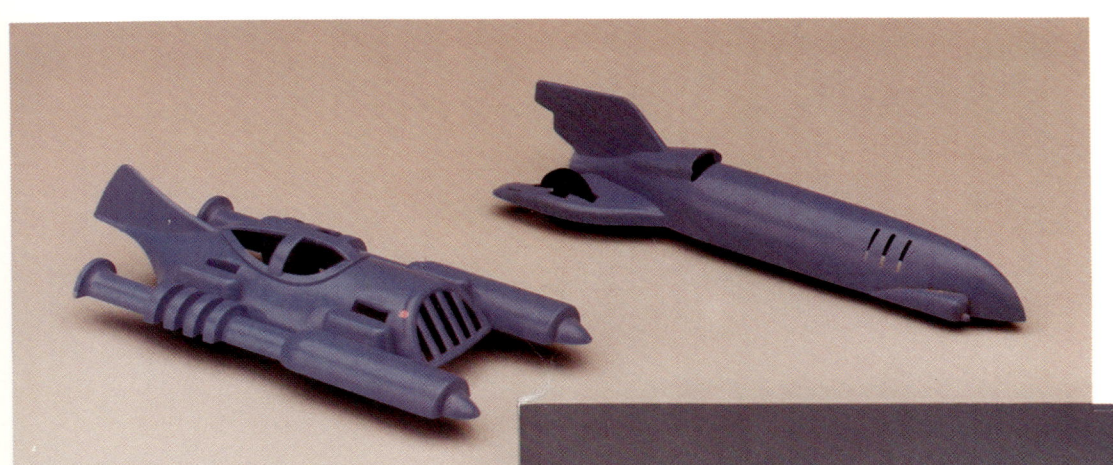

Rocket Car, 4" long, Torpedo Car, 4-1/2" long, assorted colors, plastic wheels, both part of Novelty Car Assortment, one gross per carton, Ideal Toy Corporation, USA, (No. 3013), 1953-1954. Suggested Retail $0.05 each.

Turbo Jet and Rocket Launching Platform; Turbo Jet 11-1/4" x 5-1/4" x 4", Launching Platform 9-3/4" x 4-1/2" x 4-1/2"; when crank on launching platform is turned and button is pushed, turbo jet races across the floor with siren screaming, Ideal Toy Corporation, USA (No. 4867), 1955-1958. Suggested Retail $4.00. Inspired by General Motors' 1954 Firebird I, the first gas-powered turbine car built in the United States.

Atomic Rocket Launching Truck, 12" x 5" x 5-3/4", mobile rocket launcher elevates and depresses by means of ratchet gear control, gun is loaded, cocked and fired, creating an explosion as rocket soars into space, complete with two polyetheylene cap firing rockets and opening ammo compartment to store caps, "Silvertone" finish, rubber wheels, Ideal Toy Corporation, USA, (No. 4862), 1955-1957. Renamed ICBM Launching Truck in 1958. Suggested Retail $3.00.

Satellite Launcher Truck, 17" x 6-1/2" x 6-3/4", spring powered launcher sends satellites into space, driver's cab rotates 360 degrees by means of turning dial on side, four polyethylene satellites included, assorted color combinations, Ideal Toy Corporation, USA, (No. 4874), 1957-1959. Suggested Retail $5.00.

Man From Mars, 11 1/4" tall, wind-up motor with attached key, arms move up and down as he walks, assorted colors with painted details, Irwin Corp., USA (No. 690), 1952 to mid 1950s. Suggested retail $2.00.

Aliens, 2" tall, assorted colors including metallics, Lido Toy Corporation, USA, 1952 to mid 1950s, also available in a 1-5/8" tall version. Complete set is seven different poses.

Spacemen, 2-1/4" tall, with removable helmets in two different sizes, metallic colors, same seven figures also available as soldiers in olive drab without helmets, Lido Toy Corporation, USA, 1952 to mid 1950s. Complete set is seven different poses.

SPACE TOYS 203

Rocket Car, 5-3/4" x 3-1/4" x 2", a Marx playset vehicle, with removable canopy and driver, metallic blue, plastic wheels, Louis Marx and Co., USA, 1952.

Futuristic Coupe, 3-3/4" x 1-5/16" x 1-1/4", friction motor, assorted metallic colors, rubber wheels, Louis Marx and Co., Germany, early 1950s. Futuristic Coupe, 5" x 1-3/4" x 1-3/4", assorted metallic colors and ivory, plastic wheels, Louis Marx and Co., USA, early 1950s.

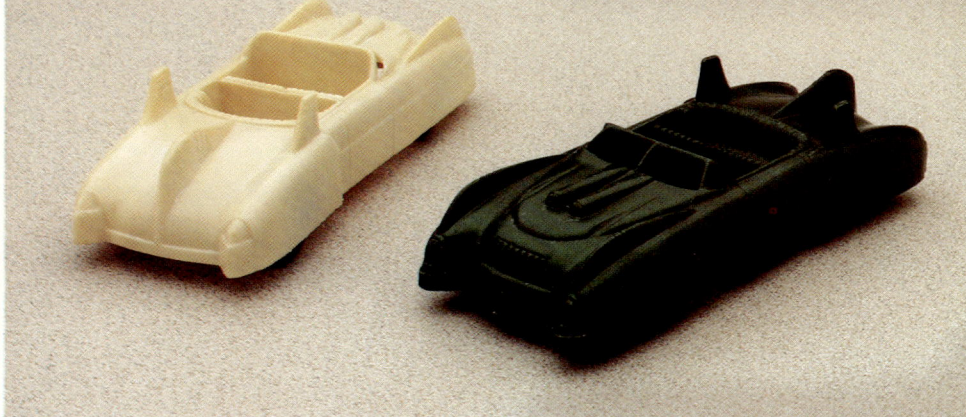

Futuristic Convertible, 5" x 1-3/4" x 1-5/16", assorted metallic colors and ivory, plastic wheels, Louis Marx and Co., USA, early 1950s.

Futuristic Convertible, 5" x 1-13/16" x 13/16", assorted metallic colors and ivory, plastic wheels, Louis Marx and Co., USA, early 1950s.

Futuristic Sedan, 5" x 1-7/8" x 1-1/4", assorted metallic colors and ivory, plastic wheels, Louis Marx and Co., USA, early 1950s.

Car of the Future, 9-3/4" x 4" x 3-1/4", with friction motor and driver, a three wheel vehicle, assorted colors, rubber wheels, Louis Marx and Co., USA, 1953 to mid 1950s.

Modern Dream Car, 10-1/2" x 5-1/4" x 3", with removable bubble top, assorted colors and plated metal front bumper, side fins and roof, rubber wheels, Mattel Inc., USA, 1953 to mid 1950s. Suggested Retail $1.99.

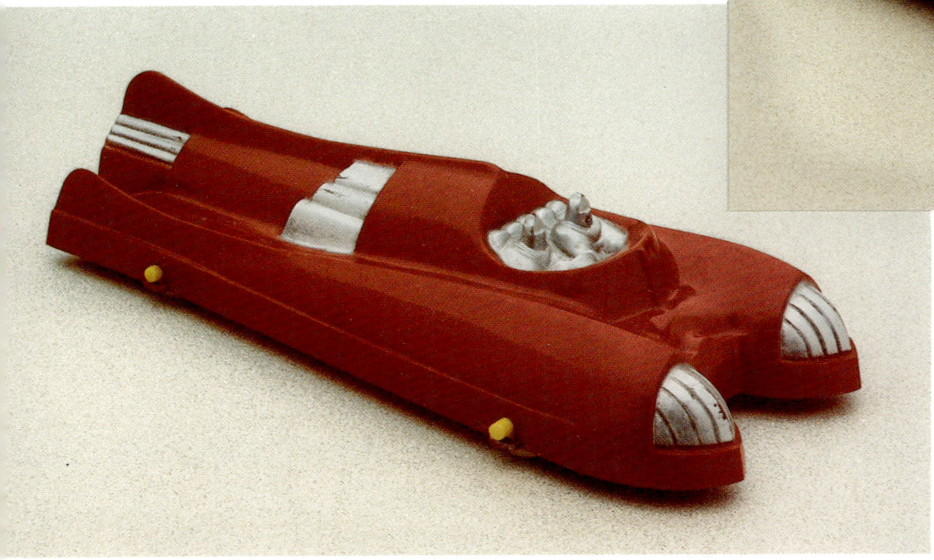

Rocket Car, 9-1/2" x 4" x 2-1/4", with driver and passenger, yellow or red with silver painted trim, plastic wheels, Plasticraft Mfg. Co., USA, early 1950s.

SPACE TOYS 205

Spacemen, 3" tall, assorted colors including metallics and vacuum metalized, Premier Products Co., USA, early to mid 1950s. Complete set is four different poses.

Rocketships, 9" long, 5" long and 3" long, two different style fuselages, assorted colors including metallics, stationary or moving plastic wheels, Premier Products Co., USA, early to mid 1950s.

Rocketship, 9" x 5" x 2", assorted colors including metallics, plastic wheels, Premier Products Co., USA, early to mid 1950s.

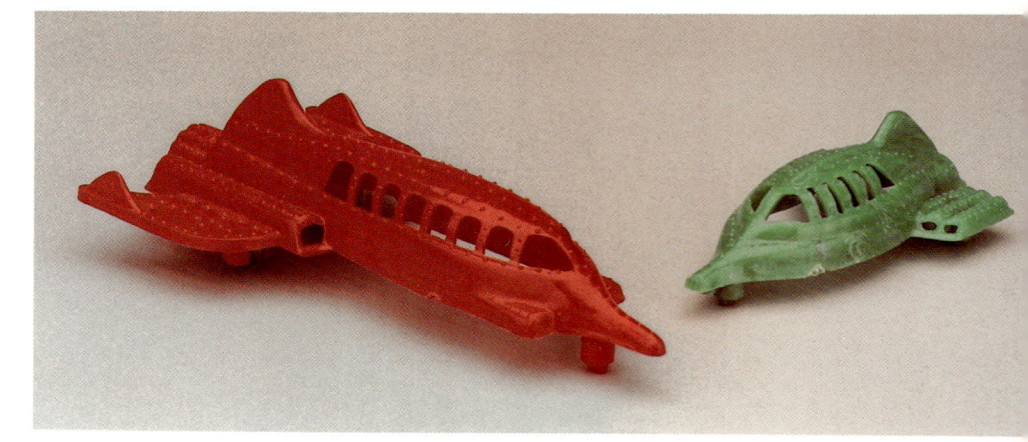

Rocketship, 5" x 2-3/8" x 1-1/4", Rocketship, 3" x 1-9/16" x 1", assorted colors including metallics, stationary or moving plastic wheels, Premier Products Co., USA, early to mid 1950s.

206 PLASTIC TOYS

Flash Gordon Space Rocket Squadron, card 12 3/4" x 11 1/2", includes; four assorted 5" long rockets and one 9" long rocket, assorted colors including metallics, plastic wheels, Premier Products Co., USA, copyright 1952 KING FEATURES SYNDICATE, INC.

X-100 Space Scout, 5" x 3-1/4" x 1-1/4", assorted colors including metallic silver, plastic wheels, Pyro Plastics Corp., USA, early 1950s.

X-200 Space Ranger, 7-1/2" x 3-1/2" x 2", assorted colors including metallic silver, plastic wheels, Pyro Plastics Corp., USA, early 1950s.

X-300 Space Cruiser, 10" x 4" x 2-1/8", with driver and removable nose cone, assorted colors including metallic silver, plastic wheels, Pyro Plastics Corp., USA, early 1950s.

SPACE TOYS 207

X-400 Space Explorer, 7-1/2" x 7" x 2", with two drivers, assorted colors including metallic silver, plastic wheels, Pyro Plastics Corp., USA, early 1950s.

Pyrotomic Energizer, 5-1/2" x 2-3/4" x 2-1/4", assorted colors including metallic silver, plastic wheels, Pyro Plastics Corp., USA, early 1950s.

Rocket Car, 3-5/8" x 1-3/4" x 1-1/8", with driver, assorted colors including metallic silver, plastic wheels, Pyro Plastics Corp., USA, early 1950s.

Pyrotomic Fire Control, 5-1/2" x 2-3/4" x 2-1/4", assorted colors including metallic silver, plastic wheels, Pyro Plastics Corp., USA, early 1950s.

Rocketship, 4-1/4" x 2" x 1-1/4", assorted colors including silver, plastic wheels, Pyro Plastics Corp., USA, early 1950s.

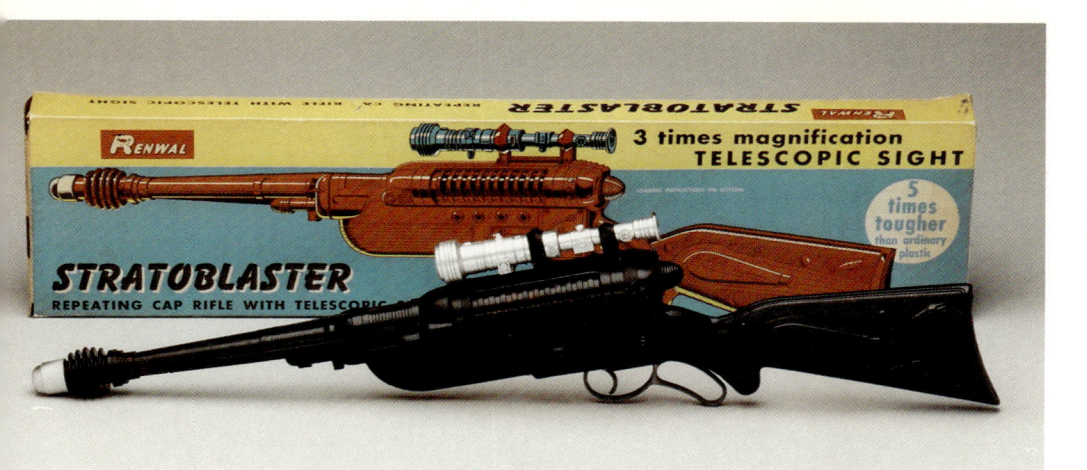

Stratoblaster, 26-3/4" long, cap shooting rifle with 3x telescopic sight, metallic colors, Renwal Manufacturing Co., Inc. USA (No. 208), 1954-1955. Suggested Retail $2.98.

Easter Bunny and Halloween Witch Rocketships; Bunny, 1-3/4" x 1-3/4" x 4", pink with blue painted details, plastic wheels; Witch, 4" x 1" x 2-3/4", orange with black painted details, plastic wheels, Manufacturer unknown, USA, early 1950s.

Take-Apart Space Man, 5-1/8" high, eight pieces to put together and take-apart again and again without gluing, assorted color combinations, Renwal Manufacturing Co., Inc. USA (No. 197), 1954.

Futuristic Race Car, 5 1/2" x 1 1/2" x 2", assorted colors, plastic wheels, Manufacturer Unknown, USA, early 1950s.

SPACE TOYS 209

CHAPTER 13
MILITARY TOYS

The number of military toys available from 1941 to 1945 was limited by World War II government restrictions on plastic, and by the fact that injection molding was still in its infancy.

Only Dillon Beck, Acme, Ideal, Thomas Toy and Renwal produced military toys during this period and their combined offerings included only boats, airplanes and jeeps. These particular types of toys are covered in length elsewhere in this book and will be discussed only briefly here.

Christmas, 1945 was finally a Christmas of peace on earth. The peace, however, came too late in the year for most molders to add any new toys to their lines.

The boats, airplanes and jeeps offered that Christmas were produced in molds made prior to or during the closing months of the war. They would help fill the toy shelves that first postwar Christmas but most Americans were war-weary and looking forward to gentler times and toys.

After doing without for four years, toy sales increased dramatically in the years immediately following the war. While dollar volume was up, more toys were sold for less money per toy, creating a tremendous demand for inexpensive toys to fill the ten to twenty-nine cent toy counter. Plastic toys with their combination of low price, play value and close resemblance to the real thing, would establish themselves early on as the leader in the inexpensive toy category.

To accomplish this, manufacturers turned out hundreds of new toys from 1946 to 1950, each requiring new tooling which could run as high as ten thousand dollars or more. Before allocating such a large sum of money, the projected sales and the potential shelf life of each new toy had to be carefully considered. Had manufacturers been able to gaze into a crystal ball and see that America would be involved in another major war in less than five years, perhaps a few more would have added military items to their lines.

There was no crystal ball, however, and with the exception of jeeps, a few propeller-driven war planes and an amphibious duck, most manufacturers shied away from any new military toys until 1951.

On June 25, 1950, Communist-ruled North Korea invaded South Korea, starting the Korean War. It was the first war in which the United Nations would play a military role and sixteen member nations sent troops to aide South Korea with the United States contributing more than ninety percent.

With approximately 480,000 American troops involved, once again there was a demand for military toys.

The first plastic military toys to reach the nation's toy stores were nothing more than civilian vehicles molded in olive drab and decorated with a white star or some other military insignia. Thus, the family sedan became an Army staff car and the familiar bulldozer became a powerful tool of the Army Corps of Engineers.

While these examples may not seem very imaginative, the fact that they were on toy shop shelves in a matter of months after the conflict started, was a dramatic demonstration of the flexibility of injection molded plastic toys compared to their metal counterparts.

Not knowing how long the conflict would last, manufacturers were again faced with difficult and expensive tooling decisions. While some, namely Ideal, Marx and Pyro, offered extensive new military lines, others seemed content to add one or two items or continue converting civilian vehicles.

The most prolific molder of military toys during this period was the Pyro Plastics Corp. of Union, New Jersey. Pyro's impressive and varied array of military hardware filled the shelves of five and dimes, drug chains and department stores across the nation.

William M. Lester started Pyro in December, 1939 after leaving Commonwealth Plastics which he helped start in 1935. Lester and his wife, Betty, would serve as president and vice president, respectively, throughout Pyro's history, until it was sold in 1971.

The name Pyro comes from the Greek word *pyr*, which means fire. The Lester's interpretation was that molded plastic objects were formed by fire (heat).

As is true of most of the companies represented in this book, Pyro was first a custom molder, and second a

U.S. Army Wrecker Truck with Ladders, 5" x 2-1/8" x 2-3/8", plastic boom with wire tow hook and two removable ladders, olive drab with pressure sensitive decals, rubber wheels, Thomas Manufacturing Corp., USA (No. 189), 1952. U.S. Army Staff Car, 4-3/8" x 1-5/8" x 1-1/4", olive drab, plastic wheels, Banner Plastics Corp., USA, early 1950s.

Sheriff Pistol, 5-1/2" long, pull trigger for clicking sound, assorted colors, Pyro Plastics Corp., USA, 1940-1949.

manufacturer of toys. The tremendous volume of toys produced by some of these companies during the late 1940s and 1950s often overshadows the fact that toys were only part of their business.

Pyro's prewar offerings included several plastic clicker type pistols and a plastic whistle, all of which are very scarce today.

After the war, Pyro's first big hit was a ferryboat with four autos that rolled on land or floated in water, followed by a working river dredge, a miniature version of Noah's ark and an extensive line of small trucks.

Like many other molders, Pyro would later try its hand at hobby kits and be quite successful.

Pyro's military and space toys produced during the early 1950s, however, are the most desirable and sought after by collectors.

In an effort to make their military toys stand out in a "sea of Pyro olive drab" Ideal also offered most of their vehicles in "Silvertone", a metallic silver plastic and "Silver Star", a vacuum metalized chrome-like finish. Judging by the number of remaining examples, olive drab was still the color of choice for most play room generals.

The Korean War ended in 1953 and by the end of 1954 most manufacturers had phased out their military lines.

Many of the military toys produced during this period can still be found. They offer plastic collectors of today a chance to add some truly inventive and historically significant items to their collection.

Personal favorites of this author are the U.S. Army Lift Truck and Railroad Handcar by Bonnie Bilt and the "U.S. Atomic Commission Mobile Unit" house trailer and sedan by the Thomas Mfg. Corporation.

By the end of 1953, space and western toys had replaced most military toys on the nation's toy shelves and an important part of American history drew to a close.

General Patton Tank, in olive drab, "Silvertone" and "Silver Star", Ideal Toy Corp., USA, 1951-1955.

U.S. Army Atomic Commission Mobile Unit, sedan 4-5/8" x 1-3/8" x 1-1/2", trailer 5-1/2" x 1-7/8" x 2", sedan has searchlight, siren and vinyl antenna, olive drab with white hot stamping and pressure sensitive decal, rubber wheels, Thomas Manufacturing Corp., USA, 1952.

Soldiers, 2-1/2" to 3" tall, all known variations shown, olive drab, Ajax Plastics, USA, early 1950s.

U.S. Army Steam Roller, 3-3/4" x 1-5/8" x 1-5/8", olive drab, plastic wheels, Banner Plastics Corp., USA, early 1950s.

U.S. Army Station Wagon, 4-3/8" x 1-5/8" x 1-3/8", olive drab, plastic wheels, Banner Plastics Corp., USA, early 1950s.

U.S. Army Tractor, 3-1/8" x 1-5/8" x 1-7/8", olive drab, plastic wheels, Banner Plastics Corp., USA, early 1950s.

U.S. Army Road Grader, 4-1/2" x 1-5/8" x 1-1/4", and Earth Scraper, 4-1/2" x 1-5/8" x 1", olive drab, rubber wheels, Banner Plastics Corp., USA, early 1950s.

U.S. Army Transport Truck with Cannon, truck 6-1/4" x 2-1/8" x 3", cannon 3-1/2" x 1-1/4" x 1-5/8", with rubber band firing mechanism and twelve plastic shells, olive drab with lithographed tin cover, rubber wheels, Banner Plastics Corp., USA (No. 218), early 1950s.

U.S. Army Ambulance, 6-1/4" x 2-1/8" x 3", with lithographed cardboard seated medic and nurse which may be seated inside, olive drab with lithographed tin cover, rubber wheels, Banner Plastics Corp., USA (No. 219), early 1950s.

U.S. Army Anti-Aircraft Gun, 4-3/4" x 2-3/4" x 3", with rubber band firing mechanism and gun that revolves and elevates, twelve plastic shells, two lithographed cardboard seated soldiers, olive drab with painted tin gun barrel, rubber wheels, Banner Plastics Corp., USA, early 1950s.

MILITARY TOYS 213

U.S. Army Comical Armored Car, 4-1/2" x 1-3/8" x 2"; when armored car is pushed along, the driver and gun go up and down and in and out respectively, olive drab with white hot stamping, plastic wheels, Bonnie Bilt, USA, early 1950s. Suggested Retail $0.25.

U.S. Army Lift Truck, 4-3/4" x 2" x 4", when crank is turned, blades go up and down realistically, olive drab with white hot stamping and assorted color combinations, rubber wheels, Bonnie Bilt, USA, early 1950s.

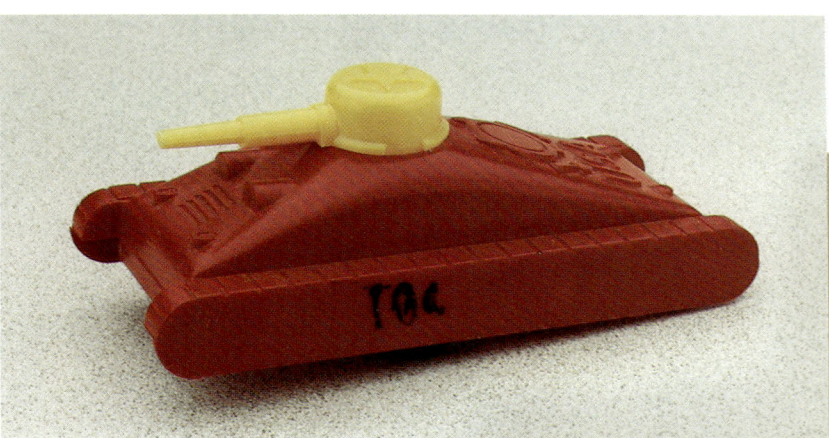

U.S. Army Railroad Handcar, 5-1/2" x 3" x 4-1/2", when pushed, riders go up and down as though they were operating the handcar, olive drab with white hot stamping and assorted color combinations, rubber wheels, Bonnie Bilt, USA, early 1950s.

USA Tank, 3-3/4" x 1-5/8" x 1-1/2", with revolving gun, assorted color combinations, plastic wheels, Dillon-Beck Manufacturing Co., USA, 1951 to mid 1950s.

U.S. Army Sherman Tank, 3-1/4" x 1-1/2" x 1-1/2", gun revolves, olive drab with white hot stamping, concealed plastic wheels, Gilmark Merchandise Corp., USA, 1951 to mid 1950s. Suggested Retail $0.10.

214 PLASTIC TOYS

U.S. Army Howitzer Carrier, 3-1/4" x 1-1/2" x 1-1/2", howitzer and gunner move left and right, olive drab with white hot stamping, concealed plastic wheels, Gilmark Merchandise Corp., USA, 1951 to mid 1950s. Suggested Retail $0.10.

U.S. Army Armored Troop Carrier, 3-1/4" x 1-3/8" x 1-1/2", with driver; gunner and gun move left and right, olive drab with white hot stamping, plastic wheels, Gilmark Merchandise Corp., USA, 1951 to mid 1950s. Suggested Retail $0.10.

U.S. Army Jeep, 3-1/4" x 1-1/2" x 1-1/2", with driver and spare wheel, olive drab, plastic wheels, Gilmark Merchandise Corp., USA, 1951 to mid 1950s. Suggested Retail $0.10.

U.S. Army Armored Car, 3-1/4" x 1-1/2" x 1-5/8", gunner and gun revolve, olive drab with white hot stamping, plastic wheels, Gilmark Merchandise Corp., USA, 1951 to mid 1950s. Suggested Retail $0.10.

U.S. Army Amphibious Duck, 5-1/4" x 2" x 1-5/8", olive drab and yellow, rubber wheels, Hubley Manufacturing Co., USA, mid to late 1950s. Suggested Retail $0.49.

MILITARY TOYS 215

Jeep, 4" x 1-1/2" x 1-1/2", assorted colors with decal, plastic wheels, Ideal Novelty and Toy Co., USA, (No. J-1), 1944-1947.

Military Police Car, 6-1/4" x 2-1/2" x 2-3/8", friction motor with siren, olive drab with white star and insignia, rubber tires on plastic hubs, Ideal Toy Corporation, USA (No. 3377), 1951-1952. Suggested Retail $0.90.

U.S. Medical Department Ambulance, 6-1/4" x 2" x 2-1/4", back doors open and close, olive drab with white or red cross insignia and decal, rubber wheels, Ideal Toy Corporation, USA (No. 4901), 1951-1952, in "Silvertone" (No. 4900), 1951-1953. Suggested Retail $0.35.

Military Engineers Bull Dozer, 6-1/2" x 3-3/4" x 2-1/2", with driver, plough moves up and down, olive drab with decal, concealed plastic wheels, Ideal Toy Corporation, USA (No. 4903), 1951-1952, in "Silvertone" (No. 4902), 1951-1953. Suggested Retail $0.50.

216 PLASTIC TOYS

U.S. Mobile Canteen Truck, 5" x 2-1/4" x 2-1/2", with driver, back doors open and close, side doors slide open, olive drab with white star and decal, rubber wheels, Ideal Toy Corporation, USA (No. 4904), 1951-1952, in "Silvertone", 1951-1953. Suggested Retail $0.35.

U.S. Air Corps Crash Boat, 5-7/8" x 1-3/4" x 1-5/8", olive drab with white insignia, Ideal Toy Corporation, USA (No. 4906), 1951-1952, in "Silvertone" (No. 4907), 1951-1952. Suggested Retail $0.15.

Army Half Track Personnel Carrier, 9-1/4" x 3" x 2-7/8", olive drab, rubber front wheels, concealed plastic rear wheels, Ideal Toy Corporation, USA (No. 4913), 1951-1952, in "Silvertone" (No. 4912), 1951-1953, in "Silver Star" (No. 4910), 1953-1955. Suggested Retail $0.80.

U.S. Army Hospital Ship, 12-1/8" x 3-1/8" x 2-1/2", olive drab with white cross and insignia, Ideal Toy Corporation, USA (No. 4915), 1951-1952; in "Silvertone" (No. 4914), 1951-1953; in "Silver Star" (No. 4921), 1953-1954. Suggested Retail $0.60.

MILITARY TOYS 217

Army Howitzer, 4-1/2" x 2" x 2", spring loaded gun actually fires, olive drab, rubber wheels, Ideal Toy Corporation, USA (No. 4916), 1951-1952; in "Silvertone" (No. 4918), 1952-1953. Suggested Retail $0.30.

U.S. Signal Corps Truck and Trailer, 6-1/2" x 2-1/4" x 2-1/4", two removable ladders, two side doors and rear door that open, 3" detachable trailer, olive drab with white star and decal, plastic wheels, Ideal Toy Corporation, USA (No. 4938), 1951-1952; in "Silvertone" (No. 4939), 1951-1953. Suggested Retail $0.50.

General Patton Tank, 6" x 3-1/2" x 3", with rotating turret and elevating gun, olive drab, concealed plastic wheels, Ideal Toy Corporation, USA (No. 4940), 1951-1952; in "Silvertone" (No. 4941), 1951-1953; in "Silver Star" (No. 4944), 1953-1955. Suggested Retail $0.70.

U.S. Army Tugboat, 6" x 1-7/8" x 1-3/4", olive drab with white insignia, Ideal Toy Corporation, USA (No. 4917), 1951; in "Silvertone" (No. 4919), 1951-1952. Suggested Retail $0.15.

U.S. Army Patrol Boat, 5-7/8" x 1-3/4" x 1-5/8", olive drab with white insignia, Ideal Toy Corporation, USA (No. 4930), 1951; in "Silvertone" (No. 4932), 1951-1952. Suggested Retail $0.15.

Army Duck, 6-3/4" x 3" x 2-3/4", with folding windshield and tow bar, olive drab, rubber wheels, Ideal Toy Corporation, USA (No. 4949), 1951-1952; in "Silvertone" (No. 4948), 1951-1953; in "Silver Star" (No. 4947), 1953-1955. Suggested Retail $0.80.

Military Vehicle Assortment, includes; Duck 3-3/4" long, Tank 3-1/4" long, Signal Corps Truck 3-1/2" long, and Jeep 3-1/4" long, assorted colors; "Silvertone" and "Silver Star" finish, plastic wheels, Ideal Toy Corporation, USA, 12 dozen per carton, (No. 3014), 1953-1954. Suggested Retail $0.05.

Great Silver Fleet, includes Duck 3-3/4" long, Wrecker 4" long, Jeep 3-3/8" long, Ambulance 3-5/8" long, Fire Truck 3-3/4" long, all in "Silver Star" finish with plastic wheels, Ideal Toy Corp., USA (No. 3045), 1954-1955. Suggested Retail $0.50. Contents will vary.

MILITARY TOYS 219

Spark Shooting Submachine Gun with Siren, 18" long, shoots a steady stream of harmless sparks when trigger housed in hand grip is pulled, replaceable flint, assorted colors, Ideal Toy Corp., USA (No. 4297), 1953-1954. Suggested Retail $2.00.

U.S. Army Stakebed Truck, 6" x 2-7/8" x 2-3/4", olive drab with white hot stamping, and assorted colors without stamping, plastic wheels, Louis Marx and Co., USA, early 1950s.

U.S. Army Jeep, 5" x 2-3/4" x 1-3/4", a Marx playset vehicle, olive drab and assorted colors, plastic wheels, Louis Marx and Co., USA, early 1950s.

Sky Sweeper Truck, 24" x 6-1/2" x 8-1/2", rotating battery powered searchlight projects six different images on wall, then twin rocket launcher fires suction cup missiles at targets; six missiles included, assorted color combinations, Ideal Toy Corporation, USA, (No. 4875), 1957-1959. Suggested Retail $8.00.

U.S. Army Troop Transport Truck, 6-3/4" x 2-7/8" x 2-1/2", a Marx playset vehicle, olive drab, plastic wheels, Louis Marx and Co., USA, early 1950s.

U.S. Army Armored Car, 6-3/4" x 3" x 2-3/4", a Marx playset vehicle, olive drab, plastic wheels, Louis Marx and Co., USA, early to mid 1950s.

U.S. Army Half Truck, 6-1/2" x 3-1/2" x 2-1/2", a Marx playset vehicle, olive drab, plastic wheels, Louis Marx and Co., USA, early to mid 1950s.

MILITARY TOYS 221

U.S. Army Antiaircraft Truck, 7-1/2" x 3-1/8" x 3-3/4", Marx playset vehicle; anti-aircraft gun and operator rotate 360 degrees and gun swivels up and down; olive drab, plastic wheels, Louis Marx and Co., USA, early to mid 1950s.

U.S. Army Duck, 7" x 3-1/8" x 3-1/2", shown alongside original factory prototype, Louis Marx and Co., USA, early 1950s.

U.S. Army Radar Truck, 7-1/2" x 3-1/8" x 4-1/2", a Marx playset vehicle; screen and operator rotate 360 degrees and radar screen swivels up and down; olive drab, plastic wheels, Louis Marx and Co., USA, early to mid 1950s.

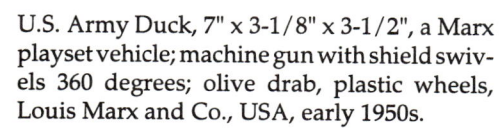

U.S. Army Duck, 7" x 3-1/8" x 3-1/2", a Marx playset vehicle; machine gun with shield swivels 360 degrees; olive drab, plastic wheels, Louis Marx and Co., USA, early 1950s.

U.S. Army Searchlight Truck, 7-1/2" x 3-1/8" x 4-1/4", a Marx playset vehicle; searchlight and operator rotate 360 degrees and light swivels up and down; searchlight has plated metal insert, olive drab, plastic wheels, Louis Marx and Co., USA, early to mid 1950s.

U.S. Army Staff Car, 3-1/2" x 1-3/8" x 1-1/4", olive drab with white hot stamping, plastic wheels, Louis Marx and Co., USA, early 1950s. Suggested Retail $0.10.

U.S. Army Ambulance, 3-1/2" x 1-3/8" x 1-1/2", olive drab with white hot stamping, plastic wheels, Louis Marx and Co., USA, early 1950s. Suggested Retail $0.10.

U.S. Army Comical Truck, 4" x 1-1/2" x 1-3/8", when truck is pushed or pulled, driver bobs up and down, olive drab with white hot stamping, plastic wheels, Plasticraft, USA, early 1950s.

U.S. Army Armored Car, 5-1/2" x 2-3/4" x 2-3/4", olive drab and other assorted color combinations, rubber wheels, Processed Plastic Co., USA, (No. 400). 1951 to mid 1950s.

MILITARY TOYS 223

Ferry Boat, 7-1/2" x 2-3/4" x 3", rolls on land and floats in water, with four 2-1/4" long sedans and/or coupes with fixed wheels, assorted color combinations, including olive drab with white hot stamping, plastic wheels, Pyro Plastics Corporation, USA, late 1940s to early 1950s.

Six U.S. Army Mobile Units, box 15-1/4" x 11-3/4" x 2-1/2", six piece set includes 7-3/4" DKW, 5-1/4" Sound Truck, 5-1/4" Radar Truck, 5-1/4" Anti-aircraft Truck, 5-1/4" Searchlight Truck, and 5-3/8" Soldier Transport, held in a red vacuum-formed tray, olive drab with white hot stamping, plastic wheels, Pyro Plastics Corporation, USA (No. 243), early 1950s.

River Dredge, 7-1/4" x 2-3/4" x 3-1/4", rolls on land and floats in water, with revolving cab and shovel that lifts by means of crank and chain, assorted color combinations, including olive drab with white hot stamping, plastic wheels, Pyro Plastics Corporation, USA, early 1950s.

U.S. Army Field Canteen, 5-3/8" x 2-1/8" x 2-3/4", olive drab with white hot stamping, rubber wheels, Pyro Plastics Corporation, USA, early 1950s.

U.S. Army Tank, 5-3/4" x 2-3/8" x 1-7/8", with gunner and revolving turret, olive drab with white hot stamping, concealed plastic wheels, Pyro Plastics Corporation, USA, early 1950s. Suggested Retail $0.39.

U.S. Army Service Truck, 5-3/8" x 2-1/8" x 2", olive drab with white hot stamping, rubber wheels, Pyro Plastics Corporation, USA, early 1950s.

U.S. Army Service Truck, 5-3/8" x 2-1/8" x 2", olive drab with white hot stamping, rubber wheels, Pyro Plastics Corporation, USA, early 1950s. Suggested Retail $0.19.

U.S. Army Soldier Transport, 5-3/8" x 2-1/8" x 2", with six seated soldiers, olive drab with white hot stamping, plastic or rubber wheels, Pyro Plastics Corporation, USA, early 1950s.

U.S. Army Gun Motor Carriage, 5" x 2-1/2" x 2-1/2", with seated figure and gun that revolves and elevates, olive drab and other assorted color combinations with and without white hot stamping, concealed plastic wheels, Pyro Plastics Corporation, USA, early 1950s.

MILITARY TOYS 225

U.S. Army DKW, 7-3/4" x 2-3/4" x 2", with removable seated driver and one standing soldier, 1-1/2" cannon and 1-1/2" jeep, olive drab and other assorted color combinations with and without white hot stamping, plastic wheels, Pyro Plastics Corporation, USA, early 1950s. When included in set (No. 243), it included two cemented seated figures and one removable standing figure only.

Twenty-One Piece U.S. Army Set, box 15-1/4" x 11-3/4" x 1-1/2"; set includes 3-3/4" Steam Roller, 3-3/8" Motorcycle, 2-1/2" Transport Wrecker, six 1-3/8" tall standing Combat Soldiers, two 2-1/4" Officer Cars, 5-1/2" Vehicle Transport Truck (trailer and cab), 5-1/2" Stake Trailer Truck (trailer and cab), 5-1/2" Transport Trailer Van (trailer and cab), and four different 4-1/8" Service Trucks, held in a red vacuum-formed tray, olive drab with white hot stamping, plastic wheels, Pyro Plastics Corporation, USA (No. 244), early 1950s.

U.S. Army Motorcycle, 3-3/8" x 1-1/4" x 1-1/2", olive drab with white hot stamping, plastic wheels, Pyro Plastics Corporation, USA, early 1950s. Suggested Retail $0.19.

U.S. Army Transport Trailer Van, 5-1/2" x 1-3/8" x 1-3/4", detachable trailer, olive drab with white hot stamping, plastic wheels, Pyro Plastics Corporation, USA, early 1950s. Suggested Retail $0.19.

U.S. Army Service Truck, 4" x 1-1/4" x 1-1/2", olive drab with white hot stamping, plastic wheels, Pyro Plastics Corporation, USA, early 1950s.

U.S. Army Stake Trailer Truck, 5-1/2" x 1-3/8" x 1-3/8", detachable trailer, olive drab with white hot stamping, plastic wheels, Pyro Plastics Corporation, USA, early 1950s. Suggested Retail $0.19.

U.S. Army Service Truck, 4-1/8" x 1-1/4" x 1-1/2", olive drab with white hot stamping, plastic wheels, Pyro Plastics Corporation, USA, early 1950s.

U.S. Army Vehicle Transport, 5-1/2" x 1-3/8" x 1-3/8", with detachable trailer carrying removable jeep and cannon or empty, olive drab with white hot stamping, plastic wheels, Pyro Plastics Corporation, USA, early 1950s. Suggested Retail $0.19.

U.S. Army Service Truck, 4-1/8" x 1-1/4" x 1-1/2", olive drab with white hot stamping, plastic wheels, Pyro Plastics Corporation, USA, early 1950s.

U.S. Army Service Truck, 4-1/8" x 1-1/4" x 1-1/2", olive drab with white hot stamping, plastic wheels, Pyro Plastics Corporation, USA, early 1950s.

U.S. Army Steam Roller, 3-3/4" x 1-1/2" x 1-5/8", front roller turns, olive drab with white hot stamping, plastic wheels, Pyro Plastics Corporation, USA, early 1950s. Suggested Retail $0.15.

U.S. Army Wrecker, 3-3/4" x 1-1/4" x 1-1/2", working wench with metal chain and tow hook, olive drab with white hot stamping, plastic wheels, Pyro Plastics Corporation, USA, early 1950s.

Super Battle Tank, 8" x 4-1/2" x 3-7/8", with friction motor and revolving turret, olive drab with decals, plastic front wheels, rubber rear wheels on plastic hubs, Saunders Tool & Die Co., USA, (No. 60). 1951 to mid 1950s. Suggested Retail $2.50.

U.S. Army Military Police, 4" x 1-3/4" x 1-1/2", with civilian driver and two passengers, folding windshield, siren and vinyl antenna, assorted colors, plastic wheels, Thomas Manufacturing Corp., USA (No. 188), 1951 to mid 1950s. Suggested Retail $0.29. Civilian drivers changed to soldiers in 1952. Available in olive drab in 1952.

U.S. Army Jeep and Trailer, 7-3/4" x 1-5/8" x 1-5/8", with civilian driver, folding windshield, permanently attached trailer with spare tire, olive drab, rubber wheels, Thomas Manufacturing Corp., USA (No. 183), 1952.

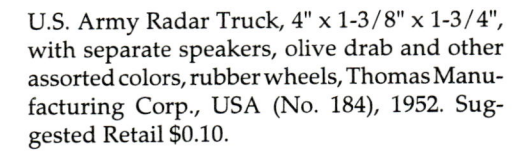

U.S. Army Radar Truck, 4" x 1-3/8" x 1-3/4", with separate speakers, olive drab and other assorted colors, rubber wheels, Thomas Manufacturing Corp., USA (No. 184), 1952. Suggested Retail $0.10.

U.S. Army Tow Truck, 4" x 1-3/8" x 1-3/4", with plastic boom, wire hook, olive drab and other assorted colors, rubber wheels, Thomas Manufacturing Corp., USA (No. 185), 1952. Suggested Retail $0.10. Staff Car, 3" x 1-5/16" x 1-3/16", olive drab and assorted colors, rubber wheels, Lido Toy Corp., USA, early 1950s.

U.S. Army Road Roller and Air Compressor, road roller 4-1/2" x 2-3/8" x 2-3/4", air compressor 3-1/2" x 1-3/4" x 1-3/4", with driver and rubber band powered front roller that steers and turns, detailed Jaeger air compressor has two removable miniature jack hammers, olive drab with pressure sensitive decals, plastic rollers and plastic and rubber wheels on compressor, Thomas Manufacturing Corp., USA, Road Roller only (No. 196), 1952. Air Compressor, 1952 to mid 1950s.

MILITARY TOYS 229

CHAPTER 14
TRAINS AND TROLLEYS

Given the overwhelming popularity of the electric toy train, it will come as a surprise to most readers that novelty plastic toy trains have the fewest examples of any category in this book.

There appear to be two reasons for this. First, no manufacturer wanted to compete with the likes of a Lionel or American Flyer, whose electric trains and accessories were generating eight to ten million dollars in annual sales as far back as 1938.

The second reason is a bit more complex. After the Second World War psychologists and educators began to work closely with the toy industry to promote year round toy sales rather than once a year sales at Christmas. The need for year round toy sales is that children grow throughout the year and need playthings that are appropriate for each period of their development, not just at Christmas or on birthdays.

This was music to the ears of toy manufacturers and most of the toys represented in this book were designed for a specific age group.

Mom knew an inexpensive plastic automobile, airplane or tea set was the perfect gift for that four to six year old, trying his or her hardest to imitate the adult world, but did Dad? When junior was two Dad bought him his first ball and bat; when he was four Dad bought him his first electric train. So much for age appropriate toys and so much for the novelty plastic toy train – it never really had a chance!

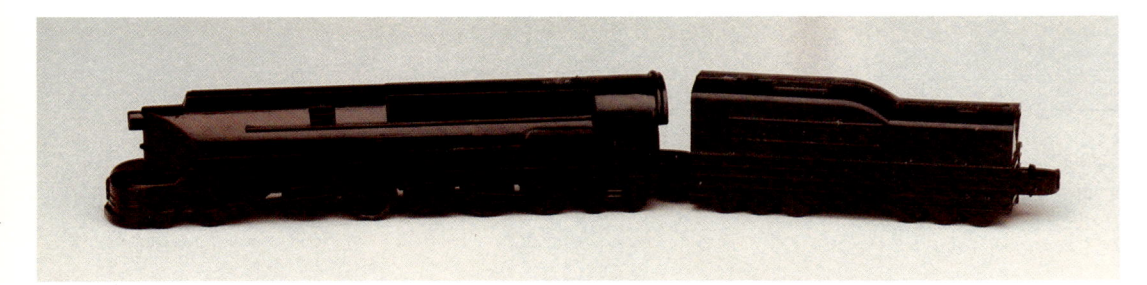

Ardee Lines Engine and Tender, engine 3-7/8" x 5/8" x 7/8", tender 2-5/8" x 5/8" x 7/8", two piece set, black, stationary plastic wheels, Ardee Plastics Co., Inc., USA, early 1950s, sold as a two piece unit in an acetate bag.

Freight Train, engine 3-3/8" x 9/16" x 1-1/8", cars 3-1/8" x 9/16" x 7/8", six piece set, assorted colors, stationary plastic wheels, Ardee Plastics Co., Inc., USA, early 1950s.

Puffo, engine 3-1/2" x 7/8" x 1-1/4", cars 3-3/4" x 3/4" x 1-1/4", six piece set, assorted colors, plastic wheels, B.W. Molded Plastic, USA, 1948 to late 1950s. Suggested Retail $0.5

Banner Fast Freight, engine 4-7/8" x 3/4" x 1-3/8", cars 4-3/4" x 7-/8" x 1-3/8", six piece set, assorted colors and vacuum metalized, plastic wheels, Banner Plastics Corp., USA (No. 155), 1948 to mid 1950s. Suggested Retail $1.19.

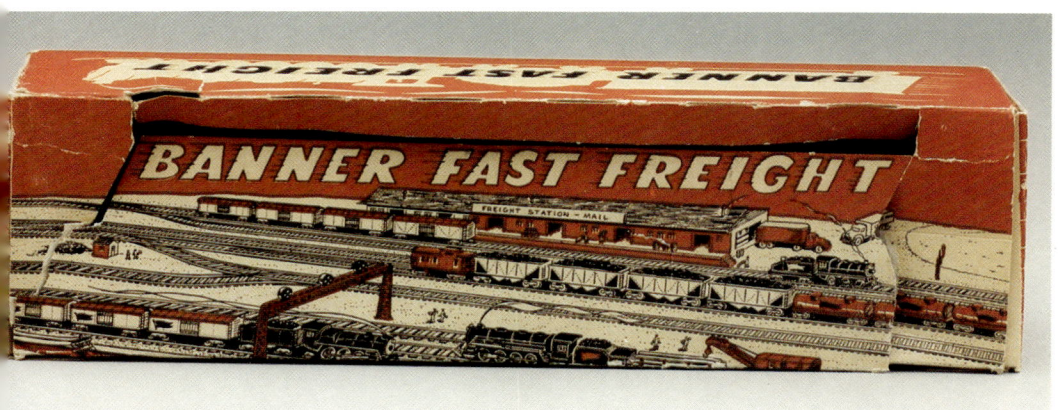

Banner Fast Freight, box 14-3/4" x 4" x 3-1/8", Banner Plastics Corp., USA (No. 155), 1948 to mid 1950s.

Freight Train, engine 2-1/8" x 5/8" x 5/8", cars 1-3/4" x 1/2" x 5/8", five piece set, assorted colors, stationary plastic wheels, Empire Plastic Corp., USA, early 1950s.

Jet Choo, 4-3/4" x 1-1/4" x 2-1/2", when balloon mounted on smokestack is blown up, locomotive zips across the floor blowing its whistle, assorted colors, concealed plastic wheels, Elmar Products Co., USA, 1951 to mid 1950s.

Pre School Locomotive, 11-3/8" x 4-5/8" x 5-1/8", bell clangs as Choochoo is pulled, assorted color combinations, plastic wheels, Ideal Novelty and Toy Co., USA (No. LO-1208), 1950, (No. 4808), 1951. Suggested Retail $1.19. Example shown missing front smokestack and light, marketed by Kleeware in England.

TRAINS AND TROLLEYS 231

Bubble Express, 12" x 5-1/4" x 8-1/2", as wheels turn, compressed air is forced through a container of bubble solution in smoke stack leaving a trail of bubbles, assorted color combinations, plastic wheels, Ideal Toy Corporation, USA (No. 4183), 1951-1954. Suggested Retail $3.00.

Diesel Locomotive, 8-3/4" x 2-1/2" x 2-1/2", four sets of wheels turn as toy is pulled along, activating realistic pistons, "Silvertone" and black, plastic wheels, Ideal Toy Corporation, USA, (No. 4810), 1953-1954. Example shown missing pistons and rear wheels.

City of Los Angeles Streamliner, engine 5-1/4" x 3/4" x 1-1/4", cars 5" x 3/4" x 1-3/16", four piece set with wind-up motor and attached key, assorted colors, plastic wheels, Nosco Plastics, USA, (No. 6321). 1948 to early 1950s. Suggested Retail $1.00. With banked track (No. 6384) 1949 to mid 1950s.

Mechanical Fast Freight, engine 3-7/8" x 7/8" x 1-1/2", cars 3-1/2" x 1" x 1-1/2", six piece set with wind-up motor and attached key, assorted colors, plastic wheels, Nosco Plastics, USA (No. 6378), 1949 to early 1950s. Suggested Retail $1.19. Available with two different type couplers during life of product and with and without a banked plastic track.

232 PLASTIC TOYS

Amusement Park Scenic Railroad, engine 4 3/8" x 1" x 2 1/2", with engineer and wind-up motor with attached key, four cars, 4 3/4" x 1 1/8" x 1 5/8", each with two removable passengers, assorted colors, plastic wheels, Nosco Plastics, USA (No. 6420), 1950 to early 1950s. Suggested retail $1.98.

Mechanical Fast Freight, box 11-1/2" x 5-1/2" x 1-1/2", Nosco Plastics, USA (No. 6378), 1949 to early 1950s.

TNT Locomotive, 4-3/4" x 1-3/4" x 3-3/8", wind-up motor with bell and bump-and-go action, black and red, metal rear wheels, concealed rubber drive wheel, Manufacturer unknown, USA, early 1950s.

TRAINS AND TROLLEYS 233

Trolley Car, 4-1/2" x 1-1/4" x 1-7/8", assorted color combinations, rubber wheels, Thomas Manufacturing Corp., USA (No. 73), 1949-1951.

Cable Trolley Car, 6-3/4" x 2-1/4" x 2-1/4", with two figures in trolley that bob up and down as trolley is pulled, assorted colors, plastic wheels, Ideal Toy Corporation, USA, (No. 3069), 1952-1953. Suggested Retail $0.40.

Tricky Trolley, 7-3/4" x 3-1/2" x 7", push trolley's pole down to wind motor, bell clangs as trolley whizzes across floor, red with painted details and lithographed paper windows, Mattel Inc., USA (No. 457), 1952 to mid 1950s. Suggested Retail $1.00.

Trolley Car, 9-1/2" x 3-3/4" x 5-3/4", when toy is pulled, bell rings and passengers bob from side to side, assorted color combinations, plastic wheels, Renwal Manufacturing Co., Inc. USA (No. 127), 1951-1953.

234 PLASTIC TOYS

CHAPTER 15
DOLL HOUSE FURNITURE

Little girls growing up in the 1940s and 50s didn't have to envy their brothers' colorful plastic toys. Many of the manufacturers represented in this book produced equally exciting lines with sister in mind. Some of the earliest applications of injection molded plastic toys were tea sets and doll house furniture.

The first injection molded plastic toy tea sets were introduced in 1942 by the Ideal Novelty and Toy Co. of Long Island City, New York. To fully appreciate this accomplishment one must realize that at the time of the introduction, plastic and the steel necessary to build the molds were both considered restricted materials.

Without the talents and contacts of Ideal's new plastics division's General Manager Islyn Thomas, plastic toy tea sets would not have been introduced until after the war. Thomas managed to secure both cellulose acetate and polystyrene for production and those sets molded during the war years alternated between the two, depending upon availability.

While plastic tea sets are not currently as popular as plastic doll house furniture with collectors, they too will undoubtedly be eagerly pursued as the interest in plastic toys increases. Boxed sets made during the war and already very scarce should be considered a welcome addition to any early plastic collection.

Plastic doll house furniture, on the other hand, already rivals most other categories presented in this book and is enthusiastically collected by both women and men.

The very first plastic doll house furniture may have been imported from England. The December, 1938 issue of *Modern Plastics* states that the English had been compression molding Bakelite phenolic plastic furniture for some time.

The first domestic injection molded doll house furniture was molded by the Wolverine Supply and Mfg. Co. of Pittsburg, Pennsylvania early in 1941.

The only sighting of Wolverine's furniture that I am aware of is in the February, 1941 issue of *Modern Plastics*. Shown in that issue are a bathtub, sink and toilet molded of Lumarith, a cellulose acetate from the Celanese Corp. of America. These three items may have been produced for less than one year as Wolverine, one of the largest lithographed metal toy makers in the United States, had converted to 100% military work by the end of 1941.

The Plastic Art Toy Corp., who marketed its toys under the PLASCO name, was molding several different sets of doll house furniture from scrap cellulose acetate during the war. The usual dark color of scrap plastic was well suited for the mahogany colored Early American furniture which moved quickly from counter to customer when available.

PLASCO apparently enjoyed a monopoly on doll house furniture throughout the war until 1945 when Renwal entered the picture.

The Renwal Manufacturing Co., Inc. of New York, New York started business in 1939 as a manufacturer of sundries. While Renwal's contribution to the war effort is unknown at this time there is no evidence of it producing any toys before 1945.

The big news for consumers was Renwal's introduction of seven plastic airplanes and five complete rooms of plastic doll house furniture in the latter part of 1945.

The timing was right and few toys have caused such a commotion upon their introduction as Renwal's colorful bathroom, bedroom, dining room, kitchen and living room doll house furniture sets.

Attractive packaging, which included a folding "full color model room" for 1946 and 1947, cheerful colors and exacting attention to detail put this miniature furniture near the top of every little girl's wish list. It was apparent from the beginning that Renwal was using "quality" plastic even though it was expensive and in short supply immediately after the war. Molded of cellulose acetate in 1945 and 1946, Renwal switched to polystyrene sometime in 1947. This eliminated the problems of warping which plagued the earlier pieces and allowed Renwal to compete against the articulated pieces being offered by Ideal.

PLASCO, whose designs seemed outdated and rather plain by comparison, continued to use scrap for many of its pieces in those early postwar years and was simply outclassed by Renwal.

Renwal quickly established itself as the major supplier of plastic doll house furniture with large displays of its wares showing up in major chain and department stores throughout the country. The Ideal Novelty and Toy Co., who had introduced a successful line of plastic tea sets in 1942, wasn't going to sit idly by and let this upstart newcomer corner another market it had its eye on. Before entering the foray, however, Ideal's first priority was to re-establish itself as America's premier doll manufacturer.

In 1947 Ideal finally introduced its line of plastic doll house furniture which included; a bathroom set, bedroom set, dining room set, garden set, two different kitchen sets, a living room set and a number of accessory pieces. The entire line appears to have been molded of polystyrene from the start with the exception of its garden trellis which was molded of cellulose acetate because of its length and difficult shape. Each boxed set came with a die-cut three-dimensional interior "beautifully illus-

Renwal Dollhouse Furniture ad, 1948.

F.W. Woolworth Co. ad, 1949.

DOLL HOUSE FURNITURE 237

trated in full color". This nice touch would be dropped in 1949 when the first window boxes would appear.

From here on out it was strictly a two horse race as Ideal and Renwal tried to one up each other, much to the delight of their multitude of fans. By 1948 Ideal's doll house furniture had replaced Renwal's in the Sears Roebuck and Co. Christmas catalog and the lead Renwal had established appeared to be narrowing.

Available as individual pieces or in boxed sets, one could easily devote an entire book to the offerings of these two companies. We will have to settle for the chronological highlights of some of their most memorable pieces.

In 1947 Renwal introduced its realistic school house with pupils and desks.

In 1947 Ideal introduced a combination television, radio and phonograph. Described in its catalog as the "piece-de-resistance", it had a television with simulated picture screen and two doors that opened to reveal a fold down phonograph with separate turntable and arm and a radio with separate knobs and dial.

Other notable Ideal offerings for 1947 included a garden pool with a blue glass insert to simulate water, a sewing machine that folded down like a real one, an ironer, and a front loading washer with opening door.

In 1948 Renwal introduced a top loading washing machine complete with removable lid, revolving agitator and working wringer. Ideal introduced a model of a real well pump that actually pumped water, an exact replica of a real lawnmower with blades that rotated when it was pushed and a working tricycle complete with miniature bell on the handelbars!

In 1949 Renwal introduced its fully jointed doll house family; mother, father, sister, brother and baby. The baby was actually introduced in 1947. Renwal also introduced an incredibly accurate model of a treadle sewing machine that not only folded down, but also had two opening drawers, a treadle that worked and a miniature metal needle that moved up and down.

Ideal introduced a sofa bed that actually converted into a bed, a folding bridge table and four chairs; they also produced a dishwasher, complete withan opening door and removable tray and dishes. Ideal also offered three different hard plastic doll house babies, two of which were fully jointed and a vinyl doll house family; mother, father, sister and brother, all with jointed arms. Only offered for one year, this family is extremely rare today. Another rare item only offered in 1949 is Ideal's "smallest baby in the world", a 1/2" long hard plastic baby in a separate bathtub!

In 1950 Renwal introduced an improved version of Ideal's poorly designed folding table and chairs and Ideal introduced its "Young Decorator" line of larger scale doll house furniture, possibly inspired by Renwal's deluxe sink, refrigerator, and stove introduced back in 1947. Ideal's sales pitch for the new line was "Little fingers need larger toys!" Available as individual pieces or in boxed sets the line included a bathroom, bedroom, dining room, kitchen, living room and nursery set.

An apparent last ditch attempt to at least capture part of the doll house furniture market, the "Young Decorator" line was never really accepted by the public and was only offered for two years. Today, examples of the "Young Decorator" line and Renwal's deluxe kitchen are much harder to find than their smaller counterparts, making them a prized addition to any collection.

By 1952 Ideal was out of the doll house furniture business. According to former Ideal President, Lionel Weintraub, Ideal could never really gain ground on Renwal, which had started producing furniture two years earlier and was well entrenched by the time Ideal entered the market.

As America's postwar generation grew up, the unprecedented demand for doll house furniture diminished. In an attempt to stimulate a declining market Renwal added intricate painted designs to almost its entire line of furniture for 1954 and 1955.

Main Street Light Pole with Opening Mailbox, 4-3/4" tall; Watering Can with Spade and Rake, 2" tall; Power Lawn Mower, 3-1/4" long, makes putt putt sound when pushed; Lovebirds in Gilded Cage, 5-3/8" tall; Wheelbarrow, 3-5/8" long; assorted colors and color combinations, Commonwealth Plastics Corp., USA, 1949 to early 1950s.

In 1956, which appears to be its final year in the furniture business, Renwal reverted back to its original look. With many of the familiar favorites dropped from the line, only six boxed sets were offered, each with fewer pieces than previous offerings.

There were, of course, other manufacturers of plastic doll house furniture or related pieces. Thomas Toy, Irwin, Commonwealth Plastics, Louis Marx and PLASCO all managed to produce a few exceptional pieces over the years. It is, however, the furniture made by Ideal and Renwal that gets collectors' blood going and will serve as a lasting tribute to the skills of America's tool and die makers.

Ad for young decorator furniture sets, 1950, Ideal Novelty & Toy Co. catalog, USA.

Kitchen Sink and Pump, 4" x 1-3/4" x 3-1/4", pumps water, missing wire soap tray and tiny dishes, yellow, Allied Molding Corp., USA, 1949 to early 1950s. Suggested Retail $0.39. Also pictured is a 2-7/8" tall vinyl girl, unknown manufacturer.

Ice Cream Cart, 4-3/4" x 2-1/8" x 2", with real bell and top that slides open, assorted color combinations, All Metal Products Co., USA, 1948 to early 1950s.

Tricycle, 4" x 2-1/2" x 2-1/2", front wheel steers by handlebars, assorted color combinations, Ideal Novelty and Toy Co., USA (No. TRI-35), 1948-1950, (No. 4893), 1951. Suggested Retail $0.35.

Play Dolls, box 9-1/4" x 7" x 2-1/4", two sitting dolls, 2-3/4" tall, two standing dolls, 3-1/2" tall, vinyl with painted details, Gerber Plastic Co., USA, 1948 to early 1950s. Suggested Retail $0.98.

240 PLASTIC TOYS

Vacuum Cleaner, 1-5/8" x 1-3/4" x 4", vinylite bag missing, (No. VC-25), 1949. Suggested Retail $0.10. Carpet Sweeper, 2-1/4" x 1-1/2" x 5-1/4", (No. CS-25), 1949. Suggested Retail $0.25. Lawn Mower, 4-1/2" x 2-1/4", blades rotate when pushed, (No. LMR-50), 1948-1949. Suggested Retail $0.49. All in assorted color combinations, Ideal Novelty and Toy Co., USA.

Well Pump, 4" x 3-3/4" x 2-3/4", actually pumps water, assorted color combinations, Ideal Novelty and Toy Co., USA (No. PMP-100), 1948-1949. Suggested Retail $0.98.

Super Market Cart, 2-1/4" x 2", with two separate baskets, assorted color combinations, Ideal Novelty and Toy Co., USA (No. SMC-10), 1950 and (No. 4252), 1951. Suggested Retail $0.10.

Hollywood Bathroom, box 10-1/2" x 8-1/2" x 3", includes; sink, water closet, hamper and Hollywood bathtub, blue with yellow accents, Ideal Novelty and Toy Co., USA (No. HB-100), 1948-1950, (No. 4420), 1951; this type of window box used 1949-1951. Suggested Retail $1.29.

DOLL HOUSE FURNITURE 241

Hamper, 2" x 1-7/8" x 7/8", (No. BH-10), 1947-1950. Suggested Retail $0.10. Bathtub, 1-1/2" x 4" x 2", (No. BT-25), 1947-1950. Suggested Retail $0.25. Wash Basin, 2-3/8" x 1-1/2" x 2", (No. BB-10), 1947-1950. Suggested Retail $0.10. Water Closet, 2" x 2" x 1-1/4", (No. BWC-25), 1947-1950. Suggested Retail $0.25. Not shown, Hollywood Bathtub, 1-1/2" x 4" x 2", (No. HBT-25), 1948-1950. Suggested Retail $0.25. All of the above sold only in sets for 1951, white and blue, Ideal Novelty and Toy Co., USA.

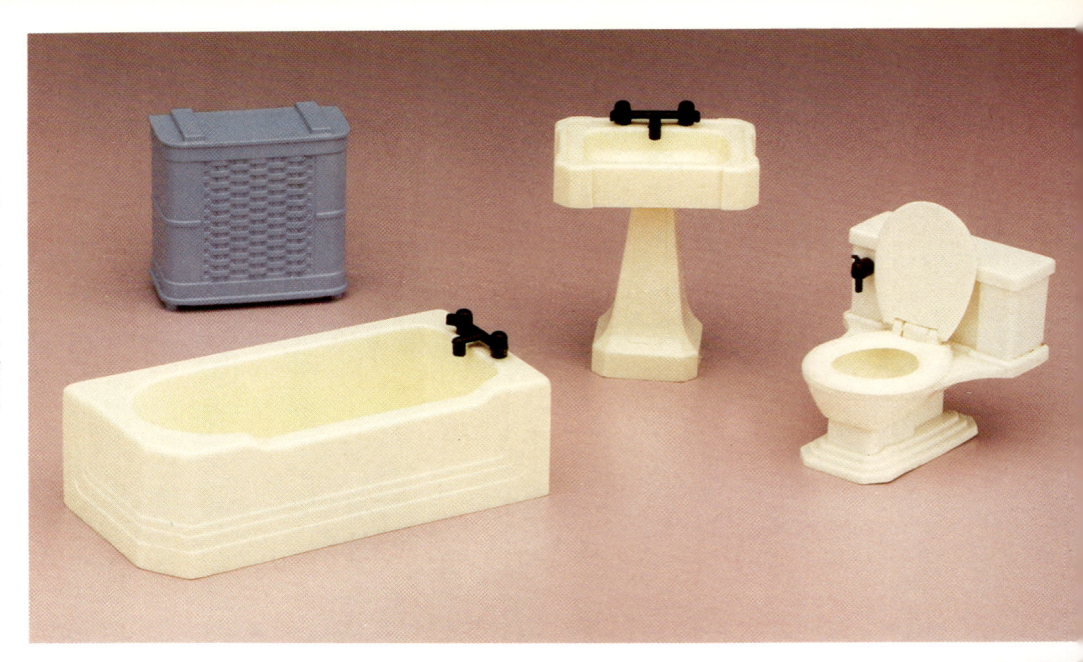

Medicine Cabinet, 2" x 2" x 1/2", with opening door and real glass mirror, two styles of handles used, white, Ideal Novelty and Toy Co., USA (No. BMC-25), 1948-1950. Suggested Retail $0.25.

Sewing Machine, 3-1/2" x 2" x 1", top opens and sewing machine automatically rises into position, mahogany and black, Ideal Novelty and Toy Co., USA (No. SM-40), 1947-1950. Suggested Retail $0.29. Sold only in a set for 1951.

One of four different box styles used in 1948, 9-1/2" x 8-1/2" x 2-1/2".

242 PLASTIC TOYS

Highboy, 4-3/8" x 2-1/2" x 1-1/8", (No. SHB-20), 1947-1950. Suggested Retail $0.19. Vanity, 3-7/8" x 3" x 1-1/4", (No. SVD-30), 1947-1950. Suggested Retail $0.29. Vanity Bench, 1-1/8" x 1-1/8" x 7/8", (No. PH-10), 1947-1950. Suggested Retail $0.10. Night Table, 1-1/2" x 1-1/8" x 1-1/4", (No. SNT-10), 1947-1950. Suggested Retail $0.10. Single Bed, 2-3/4" x 4-3/4" x 2-1/4", (No. SBE-25), 1947-1950. Suggested Retail $0.25. All of the above sold only in sets for 1951, white and blue or mahogany, Ideal Novelty and Toy Co., USA.

Folding Bridge Table and Chairs, Table 2-1/4" x 2-1/4" x 1-3/4", Chair 2-1/4" x 1" x 1", red and mahogany, Ideal Novelty and Toy Co., USA (No. BTC-70), 1949. Suggested Retail $0.69.

Hepplewhite Breakfront, 4-3/4" x 2-5/8" x 1-1/4", (No. DBR-20), 1947-1950. Suggested Retail $0.20. Hepplewhite Buffet, 2-1/4" x 4-1/8" x 1-5/8", (No. DBU-25), 1947-1950. Suggested Retail $0.25. Duncan Phyfe pieces not shown, Duncan Phyfe Arm Chair, 2-1/2" x 1-1/2" x 1-1/2", (No. DCA-10), 1947-1950. Suggested Retail $0.10. Duncan Phyfe Side Chair, 2-1/2" x 1-1/2" x 1-1/2", (No. DCH-10), 1947-1950. Suggested Retail $0.10. Duncan Phyfe Table, 2" x 4-1/2" x 2-5/8", (No. DTA-20), 1947-1950. Suggested Retail $0.19. All of the above sold only in sets for 1951, mahogany, Ideal Novelty and Toy Co., USA.

DOLL HOUSE FURNITURE

Garden Set Box Lid, box 19" x 12" x 3", (No. GSET-300), this style box used only in 1947 and 1948, Ideal Novelty and Toy Co., USA.

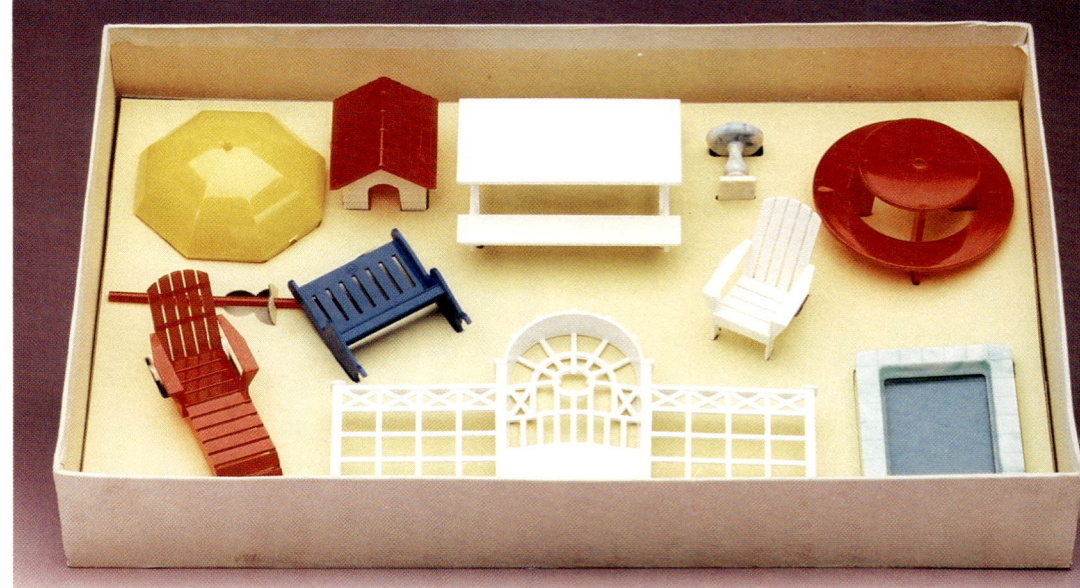

Garden Set, box 19" x 12" x 3", includes circular lawn table, trellis, lawn lounge chair, lawn bench, bird bath, lawn chair, dog house, pool, picnic table and stand up cardboard garden scene, assorted colors and color combinations, Ideal Novelty and Toy Co., USA (No. GSET-300), 1947-1948. Suggested Retail $2.98.

Lawn Chair, 2-1/2" x 2" x 2", (No. GC-20), 1947-1950. Suggested Retail $0.19. Lawn Bench, 2-3/4" x 2-1/2" x 1", (No. GB-20), 1947-1950. Suggested Retail $0.19. Trellis, 5-1/4" x 9-1/2" x 1", (No. GT-40), 1947-1948. Suggested Retail $0.39. Pool, 4-1/2" x 3" x 1/2", with blue glass, (No. GSP-35), 1947-1948. Suggested Retail $0.35. Lawn Lounge Chair, 2" x 5" x 2", (No. GA-25), 1947-1950. Suggested Retail $0.25. Lawn chair, lawn bench, pool and lawn lounge chair also sold in sets for 1951, assorted colors and color combinations, Ideal Novelty and Toy Co., USA.

Circular Lawn Table, 5-3/4" x 4-1/8" x 4-1/8", with detachable umbrella, assorted combinations of red and yellow, Ideal Novelty and Toy Co., USA (No. GTU-40), 1947-1950. Suggested Retail $0.39. Picnic Table (not shown), 2" x 3-7/8" x 4-1/2", (No. GPT-30), 1947-1950. Suggested Retail $0.29. Both sold only in sets for 1951.

Dog House, 2-3/4" x 2-1/4" x 2", (No. GDH-30), 1947-1948. Suggested Retail $0.29. Birdbath, 1-1/2" x 1-1/8" x 1-1/8", (No. GBB-10), 1947-1950. Suggested Retail $0.10. Birdbath sold only in sets for 1951, assorted colors and color combinations, Ideal Novelty and Toy Co., USA. Note: A plastic dog for the above dog house is shown in the catalog but none have ever been found, Dog Lying Down, 1-3/8" x 1/2" x 1", black, (No. GD-5), 1947-1948. Suggested Retail $0.05.

244 PLASTIC TOYS

Refrigerator, 3-1/2" x 2" x 1-3/4", (No. KF-20), 1947-1950. Suggested Retail $0.25. Range, 2-3/4" x 2-3/4" x 1-3/4", (No. KR-25), 1947-1950. Suggested Retail $0.25. Sink, 2-1/2" x 3-3/4" x 1-3/4", (No. KS-25), 1947-1950. Suggested Retail $0.25. Kitchen Table, 1-7/8" x 3-1/2" x 2-1/4", (No. KT-10), 1947-1950. Suggested Retail $0.10. Kitchen Chair, 2-1/2" x 1-1/8" x 1-1/4", (No. KC-5), 1947-1950. Suggested Retail $0.05. All of the above sold only in sets for 1951, white, with black or red accents, Ideal Novelty and Toy Co., USA.

Deluxe Kitchen includes; Cabinet (uncataloged), 2-1/2" x 1-7/8" x 5-1/4", Refrigerator, 4-1/8" x 2-1/2" x 2-1/2", with opening door, (No. KRF-40), 1947-1950. Suggested Retail $0.39. Range, 2-3/4" x 3-3/4" x 2-1/8", with opening oven and utility doors and roasting pan shown on the cabinet counter, (No. KRA-50), 1947-1950. Suggested Retail $0.49. Sink and Tub Combination, 2-3/4" x 5" x 1-7/8", with opening doors, swinging spigot and sliding drain board, (No. KST-50), 1947-1950. Suggested Retail $0.49. White with black accents, Ideal Novelty and Toy Co., USA.

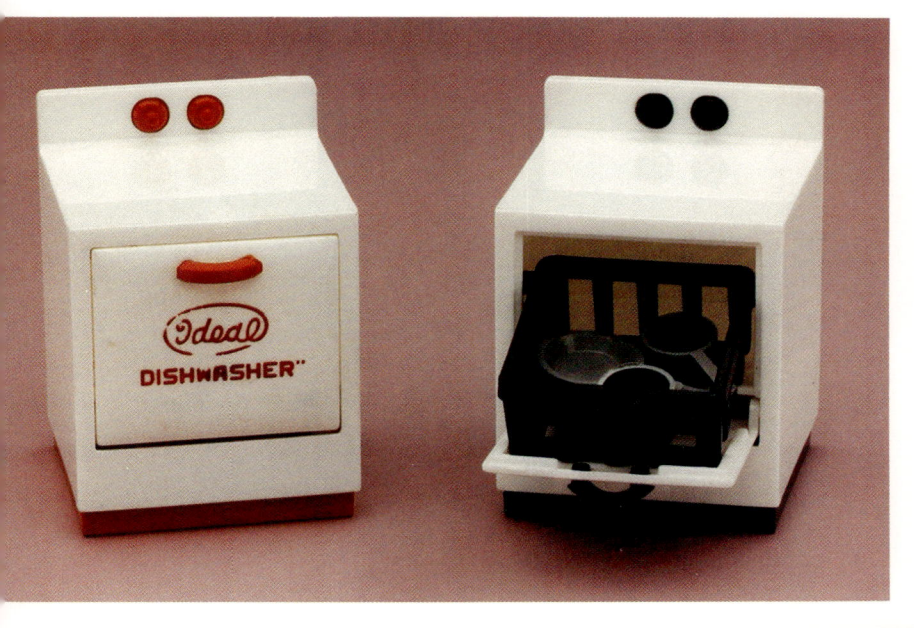

Dishwasher with Dishes, 1-7/8" x 1-7/8" x 2-5/8", with removable tray and tiny dishes, white with black or red accents, Ideal Novelty and Toy Co., USA (No. DW-25), 1949-1950. Suggested Retail $0.25.

Electric Washer, 2-7/8" x 2-1/8" x 2", replica of a real Bendix washing machine with opening door, white with black accents, (No. LW-35), 1947-1950. Suggested Retail $0.35. Electric Ironer, 2-5/8" x 2-1/2" x 1-1/2", with hinged cover that opens, white with black accents, (No. LM-35), 1947-1950. Suggested Retail $0.35. Both sold only in sets for 1951, Ideal Novelty and Toy Co., USA.

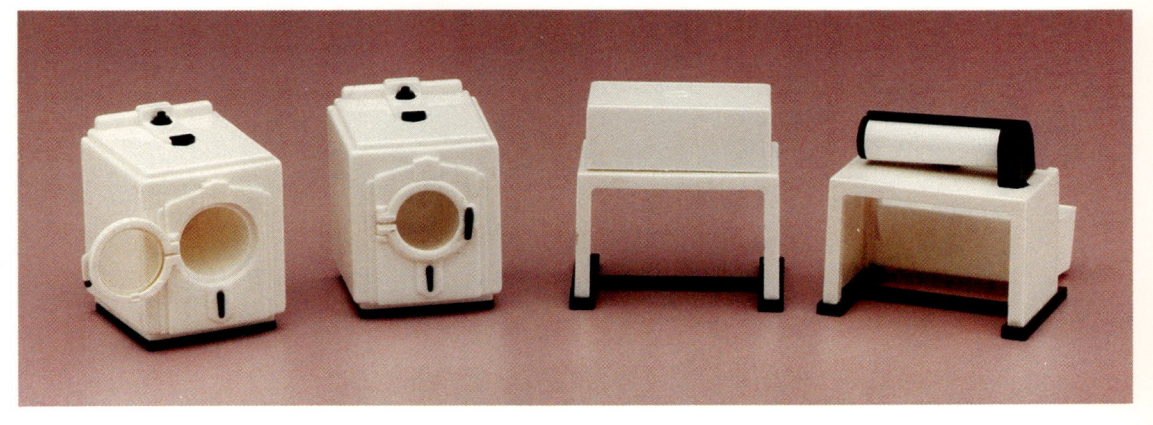

DOLL HOUSE FURNITURE 245

Living Room Set, box 10-1/2" x 5-1/2" x 6", top folds up to form the roof of the house, includes; Queen Ann Wing Chair, Queen Ann Lounge Chair, Queen Ann Sofa, Coffee Table, Floor Lamp and Table Lamp, assorted colors and color combinations, Ideal Novelty and Toy Co., USA (No. DHP-120), 1948. Suggested Retail $1.19. One to four box styles used in 1948.

Queen Ann Wing Chair, 3-1/4" x 2-1/2" x 2", (No. PWC-20), 1947-1950. Suggested Retail $0.19. Queen Ann Sofa, 2-1/2" x 4-3/4" x 1-3/4", (No. PSO-25), 1947-1950. Suggested Retail $0.25. Coffee Table, 1" x 2-1/2" x 2", (No. PCT-10), 1947-1950. Suggested Retail $0.10. Floor Lamps, 4" x 1-1/4" x 1-1/4", (No. PFL-10), 1947-1950. Suggested Retail $0.10. Queen Ann Lounge Chair, 2-1/4" x 2-1/8" x 2", (No. PLC-20), 1947-1950. Suggested Retail $0.19. Assorted colors, all of the above sold only in sets for 1951, Ideal Novelty and Toy Co., USA.

Picture Frame, 1-1/2" x 1-1/2" x 1/4", (No. PPF-10), 1948. Suggested Retail $0.10. Table Lamp, 1-3/4" x 7/8" x 7/8", (No. PTL-5), 1947-1950, mahogany, sold only in sets for 1951, Ideal Novelty and Toy Co., USA. Suggested Retail $0.05.

Cabinet Radio, 2" x 1-1/2" x 3/4", (No. PRD-10), 1947-1950. Suggested Retail $0.10. Early American Fireplace, 3" x 4-1/2" x 1-1/2", (No. PFP-40), 1947-1950. Suggested Retail $0.39. Radiator, 1-1/4" x 3-1/4" x 3/4", (No. RTR-15), 1947-1949. Suggested Retail $0.15. Mahogany and gray, Ideal Novelty and Toy Co., USA.

Combination Television Radio and Phonograph, 3" x 3-1/4" x 1-1/4", side doors open to display phonograph and radio, mahogany with yellow accents, (No. PRT-75), 1947-1950, Ideal Novelty and Toy Co., USA. Suggested Retail $0.75.

Tilt Top Pie Crust Table, 2-1/4" x 2-1/4" x 2-1/4", top tilts forward, mahogany, Ideal Novelty and Toy Co., USA (No. PTT-15), 1947-1950 and 1951 in sets only. Suggested Retail $0.15.

Piano and Bench, Piano 2-3/4" x 3-1/2" x 2-1/2", Bench 1-3/4" x 3/4" x 1", mahogany, Ideal Novelty and Toy Co., USA (No. PIA-40), 1947-1950. Suggested Retail $0.39.

Secretary, 5-3/4" x 2-3/4" x 1-1/2", desk lid opens to form writing desk, mahogany, Ideal Novelty and Toy Co., USA (No. PSY-25), 1947-1950. Suggested Retail $0.25.

DOLL HOUSE FURNITURE 247

Sofa Bed, 5-1/8" x 2-1/2" x 2-1/4", back slides down to form double sleeper, assorted colors, Ideal Novelty and Toy Co., USA (No. SSB-40), 1949. Suggested Retail $0.39.

Sofa Bed, open, Ideal Novelty and Toy Co., USA (No. SSB-40), 1949. Suggested Retail $0.39.

Nursery Set, box 10-1/2" x 5-1/2" x 6", top folds up to form the roof of the house, includes; swinging cradle, doll bath, chest of drawers, table lamp, night table and standing baby, assorted colors, Ideal Novelty and Toy Co., USA (No. DHN-120), 1948. Suggested Retail $1.19. One of four box styles used in 1948.

Plastic Drop Side Crib, 2-1/4" x 4", side drops down like real crib, (No. NCR-30), 1948-1950 and 1951 only in sets. Suggested Retail $0.29. Folding Play Pen, 3" x 3" x 3", folds flat, (No. NP-25), 1948-1950. Suggested Retail $0.25. Rocker High Chair, 3-3/4" x 2", collapses to create rocker, (No. NHR-35), 1948-1950 and 1951 in sets only. Suggested Retail $0.35. Assorted colors and color combinations, Ideal Novelty and Toy Co., USA.

248 PLASTIC TOYS

Strollmaster (Stroller-Walker), 3-1/2" x 2-3/4", handle can be folded down to convert stroller to walker, (No. NSM-40), 1948-1951, 1951 in sets only. Suggested Retail $0.39. Carriage, 3" x 4", hood moves forward and backward, (No. NCG-30), 1948-1950. Suggested Retail $0.29. Assorted color combinations, Ideal Novelty and Toy Co., USA.

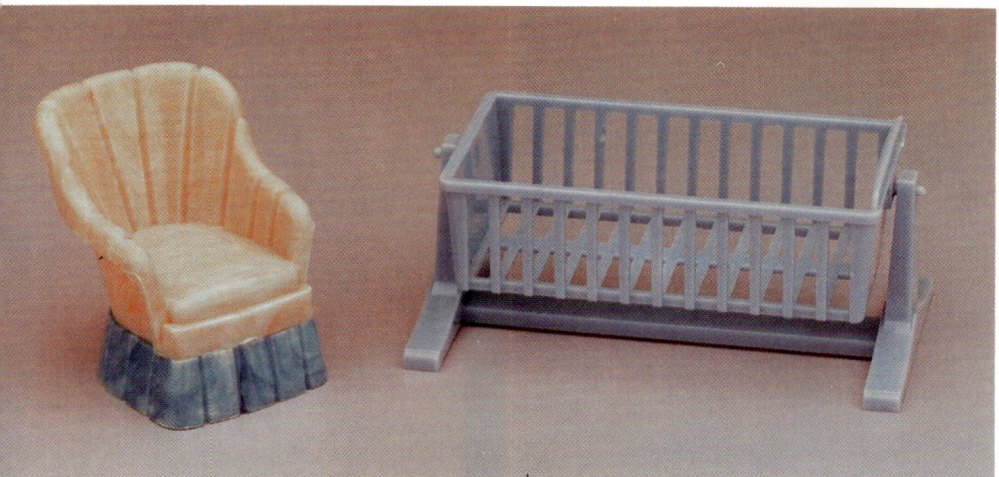

Bedroom Chair, 2-1/4" x 1-3/4" x 2", (No. SBC-15), 1947-1950. Suggested Retail $0.15. Swinging Cradle, 3-3/4" x 2-1/4" x 1-1/2", (No. NCS-20), 1948-1950. Suggested Retail $0.19. Assorted colors and color combinations, Ideal Novelty and Toy Co., USA.

Training Seat and Potty, Seat 2" x 1-3/4" x 2", Potty 1-1/8" x 1-5/16" x 1/2", potty slides under seat, assorted colors, Ideal Novelty and Toy Co., USA (No. NTC-15), 1948-1950. Suggested Retail $0.15.

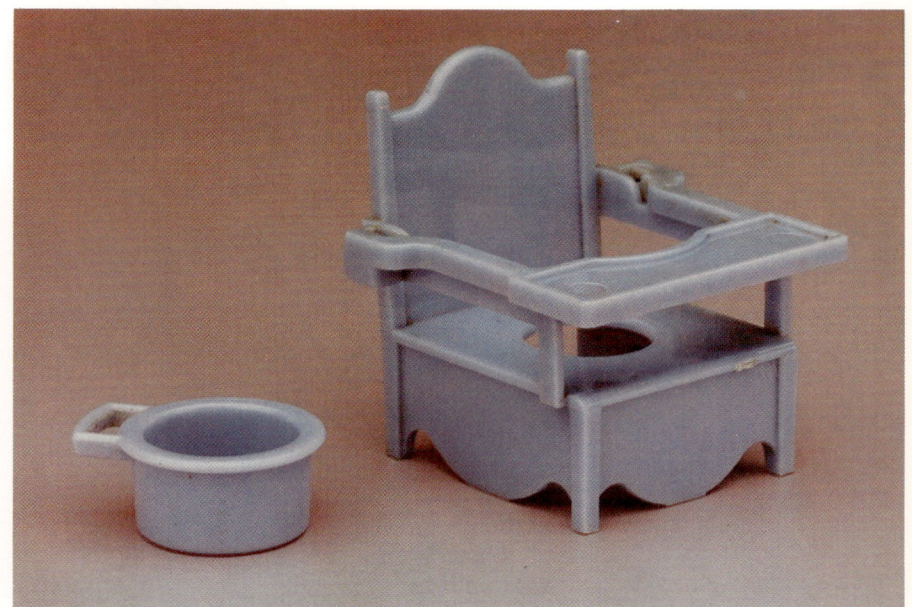

Doll Bath, 2-1/4" x 3-1/4", top folds out of way to use bath, assorted color combinations, Ideal Novelty and Toy Co., USA (No. NBA-30), 1948-1950 and 1951 in sets only. Suggested Retail $0.29.

DOLL HOUSE FURNITURE 249

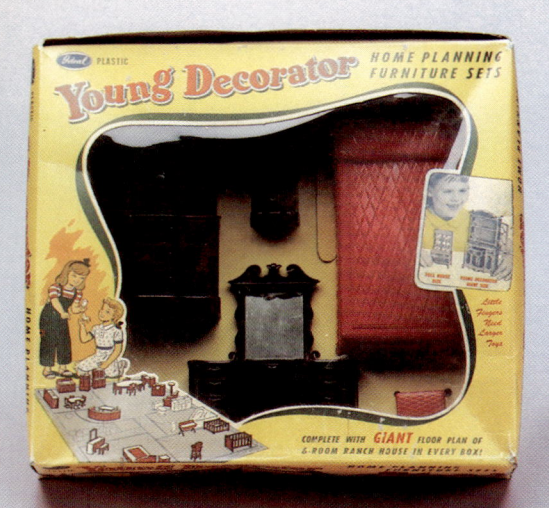

Young Decorator Bedroom Set, box 13-1/4" x 12-1/4" x 4", includes vanity dresser, vanity bench, bedroom chest, bed, and night table, mahogany and pink, Ideal Novelty and Toy Co., USA (No. JS-150), 1950, (No. 3620), 1951. Suggested Retail $1.49.

Young Decorator Box Side Panels, boxes 13-1/4" x 12-1/4" x 4", Ideal Novelty and Toy Co., USA, 1950-1951. Suggested Retail $1.49.

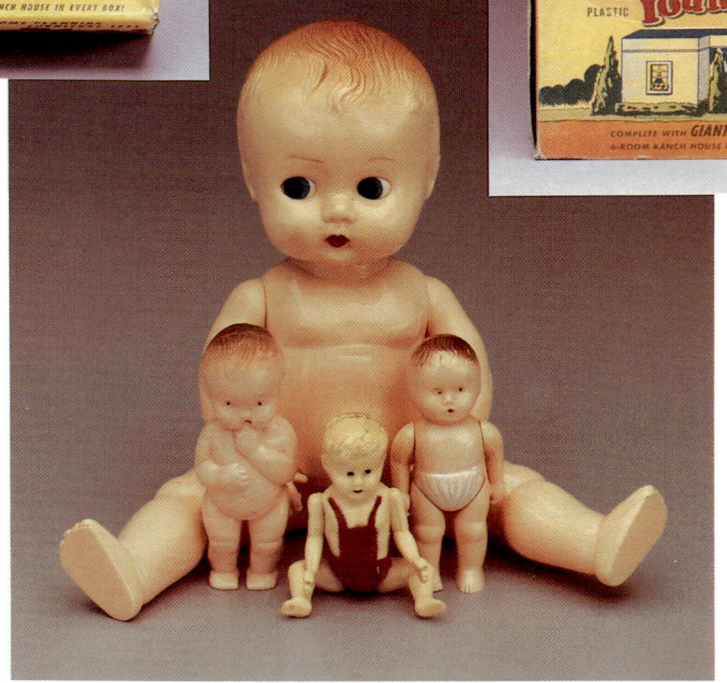

Boopsie Doll, 7-1/2" tall, fully jointed with sleeping eyes, (No. PBJ-70), 1950, (No. 3120), 1951-1952. Suggested Retail $0.69. Standing Plastic Baby Doll, 2-1/2" tall, without clothes, (No. PB-10), 1948-1949. Suggested Retail $0.10. Fully Jointed Plastic Baby Doll, 2-1/2" tall, with painted sunsuit, (No. PBS-10), early 1948 only. Suggested Retail $0.10. Fully Jointed Plastic Baby Doll, 3" tall with painted diapers, (No. PBJ-10) replaced sunsuit doll, mid 1948-1950 and 1951 in sets only, Ideal Novelty and Toy Co., USA.

Young Decorator Bathroom Set, Bathtub, 7-1/8" x 5-1/8" x 1-5/8", (No. JBT), 1950, (No. 3611), 1951; Chair, 3-3/8" high, (No. JKC), 1950, (No. 3641), 1951; Wash Stand, 5-5/8" x 1-7/8" x 3-3/8", (No. JBS), 1950, (No. 3612), 1951; Water Closet, 3" x 3" x 2-5/8", (No. JBW), 1950, (No. 3614), 1951; Waste Can, 2-1/8" high, (No. JBC), 1950, (No. 3613), blue and yellow, Ideal Novelty and Toy Co., USA.

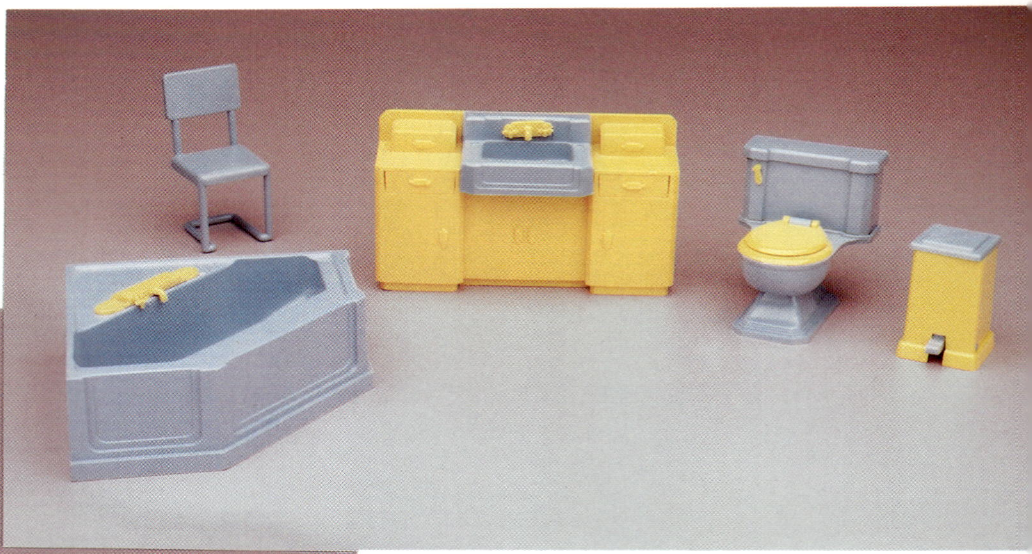

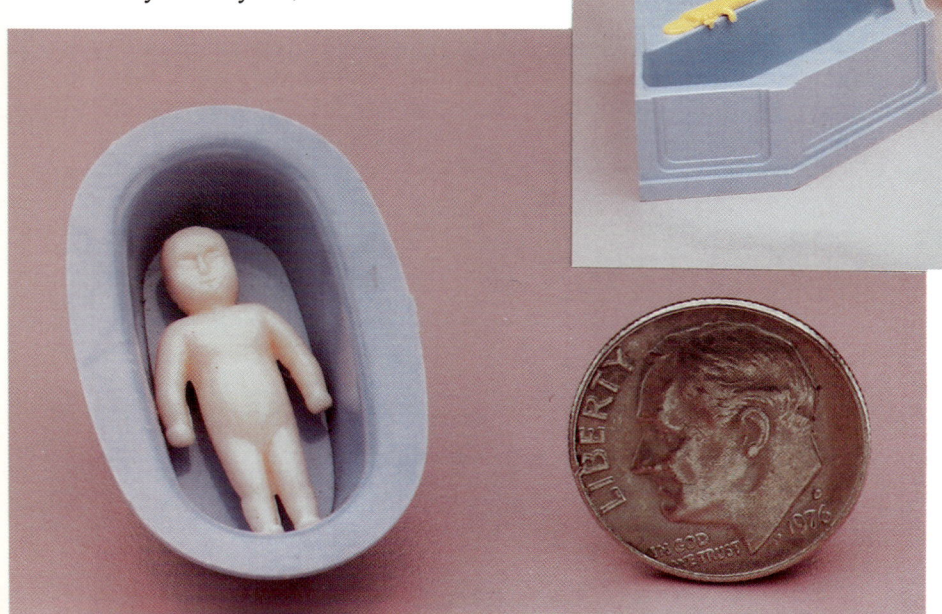

Smallest Baby in the World, 1/2" tall with separate bathtub, assorted colors, Ideal Novelty and Toy Co., USA (No. SBW-5), 1949. Suggested Retail $0.05.

Young Decorator Bedroom Set, Vanity Dresser, 6-1/2" x 4-7/8" x 2-1/8", (No. JSV), 1950, (No. 3625), 1951; Vanity Bench, 1-5/8" x 1-1/8" high, (No. JSB), 1950, (No. 3624), 1951; Night Table, 2-1/4" tall, (No. JSN), 1950, (No. 3623), 1951; Bed, 7" x 4" x 3-5/8", (No. JSX), 1950, (No. 3621), 1951; Bedroom Chest, 6-7/8" x 4-1/8" x 2-1/4", (No. JSC), 1950, (No. 3625), 1951, mahogaany and pink, Ideal Novelty and Toy Co., USA.

Young Decorator Dining Room Set, Dining Room Buffet, 6" x 3-1/4" x 1-7/8", (No. JDB), 1950, (No. 3631), 1951; Dining Room Table, 6-5/8" x 4" x 2-5/8", (No. JDT), 1950, (No. 3634), 1951; Dining Room Chair, 3-3/8" high, (No. JDC), 1950, (No. 3633), 1951; China Closet, 8" high x 4" wide, (No. JDX), 1950, (No. 3632), 1951, mahogany and yellow, Ideal Novelty and Toy Co., USA.

Young Decorator Kitchen Set, Refrigerator, 6 3/8" x 3-1/4" x 2-1/4", (No. JKF), 1950, (No. 3645), 1951; Kitchen Table, 4-1/2" x 2-3/4" x 2-7/8", (No. JKT), 1950, (No. 3644), 1951; Kitchen Chair, 3-3/8" high, (No. JKC), 1950, (No. 3641), 1951; Range, 4-1/2" x 2-1/4" x 4", (No. JKR), 1950, (No. 3642), 1951; not shown Kitchen Sink, 6" x 3-1/2" x 2-1/2", (No. JKS), 1950, (No. 3643), 1951, white with black accents, Ideal Novelty and Toy Co., USA.

Young Decorator Living Room Set, Two End Section Chairs, 2-5/8" x 2-5/8" x 2-7/8" each, (No. JPE), 1950, (No. 3654), 1951; Center Square-Section Chair, 2-1/8" x 2-5/8" x 2-7/8", (No. JPS), 1950, (No. 3652), 1951; Torchiere Lamp, 5-7/8" high; Bowl 1-7/8" dia., (No. JPL), 1950, (No. 3656), 1951; not shown, Coffee Table, 4-1/2" x 2" x 1-1/4", (No. JPX), 1950, (No. 3653), 1951, and Center Curved-Section Chair, 3-1/2" x 2-5/8" x 2-7/8", mahogany and pink, Ideal Novelty and Toy Co., USA.

DOLL HOUSE FURNITURE 251

Young Decorator Living Room Set, Television, 3-3/4" x 2" x 3-7/8", 6 simulated channels change as knob is turned, (No. JPT), 1950, (No. 3655), 1951; Doll House Family Man, 5" tall, (No. VMU-30), 1949. Suggested Retail $0.29; Not shown, Woman, 5" tall (No. VWU-30), 1949. Suggested Retail $0.29; Boy, 3" tall, (No. VBU-20), 1949. Suggested Retail $0.19. Girl, 3" tall, (No. VGU-20), 1949. Suggested Retail $0.19. All the above figures are painted vinyl with jointed arms held on by a pin, Ideal Novelty and Toy Co., USA.

Four Piece Tea Set, Tray 2-1/2" long, Irwin Corp., USA, 1948 to early 1950s.

Baby Stroller and Baby; Stroller 3-3/4" x 2-3/8" x 3", Baby, 3" long, pink and blue, Jericho Toy Mfg. Corp., USA.

Street Sweeper Push Cart and Brush, Cart, 3" x 2-1/2" x 2-3/8"; Brush, 5" long, assorted color combinations; Doll on Toilet, Doll, 2-5/8" tall, pink with painted detail; Toilet, 2-1/16" tall, lid opens, white and red, Irwin Corp., USA, late 1940s to early 1950s.

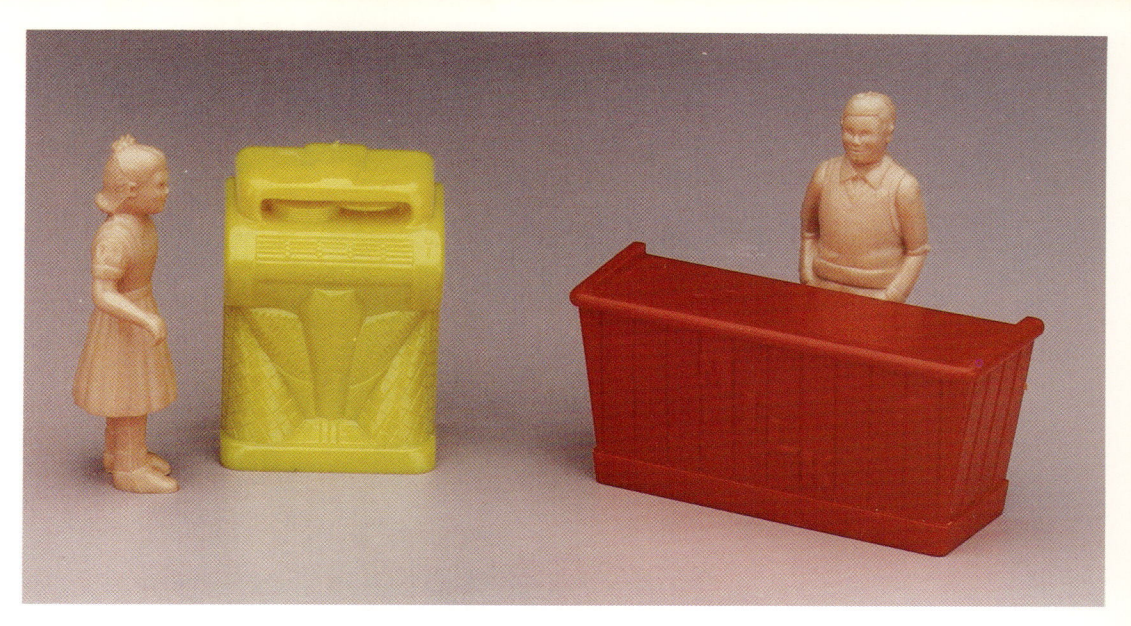

Jukebox, 2-1/8" tall; Milk Bar, 3-1/8" long; Vinyl Sister Doll, 2-1/2" tall; Vinyl Brother Doll, 2-3/4" tall, typical accessories included in Marx lithographed metal doll houses, Louis Marx and Co., Inc., USA, early 1950s.

Miniature Utensils, 3/4" to 2" long, toaster, rolling pin, waffle iron, mixer and bowl, radio and bread on cutting board with knife, assorted colors, Louis Marx and Co., Inc., USA, 1956.

Folding Ladder, 3-7/8" tall, folds like real ladder, assorted colors, Louis Marx and Co., Inc., USA, early 1950s.

DOLL HOUSE FURNITURE 253

Small Laundry Room Furniture, Sink, 2-1/16" x 1" x 1-3/16", typical of furniture included in Marx lithographed metal doll houses, Louis Marx and Co., Inc., USA, early to mid 1950s.

Large Laundry Room Furniture, Washing Machine, 2" tall, white, typical of furniture included in Marx lithographed metal doll houses, (early Marx Vinyl Doll House Doll's Mother, 3" tall, Sister, 2-1/2" tall), Louis Marx and Co., Inc., USA, early to mid 1950s.

Patio Furniture, Table with Umbrella, 2-7/8" tall, assorted colors and color combinations, typical of furniture included in Marx lithographed metal doll houses, Louis Marx and Co., Inc., USA, early to mid 1950s.

Doll House Accessories, Telephone, 1-1/8" x 7/8" x 5/8", (No. 28), 1949-1953. Suggested Retail $0.05; Table Radio, 1-1/4" x 5/8" x 5/8", (No. 16), 1948-1952. Suggested Retail $0.05. Scale, 7/16" x 1" x 5/16", (No. 10), 1948-1953. Suggested Retail $0.05. Baby Scale, part of Hospital Nursery Set, (No. 214), 1954-1956, Stool, 1-1/4" x 7/8", (No. 12), 1948-1952. Suggested Retail $0.05. Smoking Stand, 1-15/16" x 11/16", (No. 13), 1948-1953. Suggested Retail $0.05. Mantle Clock, 11/16" x 1-9/16" x 1/2", (No. 14), 1948-1953. Suggested Retail $0.05. Step-On-Can and Dust Pan, Can, 1-1/8" x 1-1/4" x 3/4", Pan, 1-1/4" x 7/8" x 1/4", (No. 64), 1949-1953. Suggested Retail $0.10. Kitchen Clock, 11/16" x 5/8" x 5/16", (No. 11), 1948-1953. Suggested Retail $0.05. All came in assorted colors and color combinations, Renwal Mfg. Co., Inc., USA.

Bathtub, 1-5/8" x 2-1/8" x 4", (No. 95), 1945-1953 and 1956. Suggested Retail $0.25. Bathroom Sink, 2-7/16" x 2" x 1-5/16", (No. 96), 1945-1953 and 1956. Suggested Retail $0.10. Toilet, 2-3/16" x 1-11/16" x 2-1/8", (No. 97), 1945-1953 and 1956. Suggested Retail $0.25. Hamper, (No. 98), 1945-1953 and 1956. Suggested Retail $0.10. Assorted colors, Renwal Mfg. Co., Inc., USA.

Server, 2-1/4" x 3-1/2" x 1-5/8", (No. 54), 1945-1953 and 1956. Suggested Retail $0.19. Dining Table, (No. 51), 2" x 4-1/8" x 2-5/8", 1945-1953 and 1956. Suggested Retail $0.25. Chair, 2-5/8" x 1-1/4" x 1-3/8", (No. 53), 1945-1953 and 1956. Suggested Retail $0.05. China Closet, 3-3/4" x 2-13/16" x 1-1/8", (No. 52), 1945-1953. Suggested Retail $0.25. Not shown, Buffet, (No. 55), 1945-1951. Assorted colors and color combinations, Renwal Mfg. Co., Inc., USA.

Highboy, 3-3/16" x 2-5/16" x 1-1/16", (No. 85), 1945-1953 and 1956. Night Table, 1-1/2" x 7/8" x 1", (No. 84), 1945-1953 and 1956. Suggested Retail $0.05. Bed, 2-1/4" x 2-1/4" x 3-15/16", (No. 81), 1945-1953 and 1956. Suggested Retail $0.25. Vanity, 4-1/16" x 2-7/16" x 1-1/4", (No. 82), 1945-1953. Suggested Retail $0.25. Vanity Bench, 1-1/4" x 7/8" x 2", (No. 75), 1945-1953. Suggested Retail $0.05. Not shown, Dresser, 4-5/16" x 2-7/16" x 1-5/16", (No. 83), 1945-1953 and 1956. Suggested Retail $0.25. All came in assorted colors, Renwal Mfg. Co., Inc., USA.

Bridge Set, box 7" x 7" x 2-1/2", assorted color combinations, Renwal Mfg. Co., Inc., USA (No. 1089), 1951-1955.

Vacuum Cleaner, 4" x 1-1/2" x 1-1/4", assorted color combinations with decal on vinyl bag, (No. 37), 1949-1953. Suggested Retail $0.10. Carpet Sweeper, 4-1/4" x 1-1/4" x 3/8", rollers turn, assorted color combinations, (No. 116), 1950-1952. Suggested Retail $0.05. Not shown, Mop with cotton yarn head, 4-7/8", (No. 117), 1950-1952. Suggested Retail $0.10. Broom with bristle head, 4-1/2", (No. 121), 1950-1952. Suggested Batrail $0.10. Renwal Mfg. Co., Inc., USA.

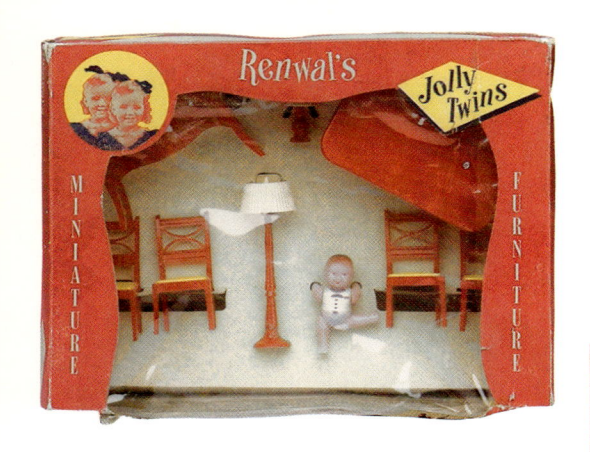

Dining Room Set, box 10-1/2" x 8" x 3", includes server, table, four chairs, floor lamp, table lamp and baby with painted clothes, assorted colors and color combinations, this set was apparently made with different combinations of furniture, Renwal Mfg. Co., Inc., USA (No. 912), 1949-1953. Suggested Retail $0.98

Kitchen Set, box 12-1/4" x 8-3/4" x 2-1/2", includes sink, stove, refrigerator, table and four chairs, and fold out lithographed cardboard kitchen, white and black, Renwal Mfg. Co., Inc., USA (No. unknown), 1945-1947. Suggested Retail $0.98.

Stove, 2-5/8" x 2-7/8" x 2", (No. 69), 1945-1947. Suggested Retail $0.25. Refrigerator, 3-1/2" x 1-15/16" x 1-5/16", (No. 66), 1945-1947. Suggested Retail $0.25. Kitchen Sink, 2-11/16" x 3" x 2", (No. 68), 1945-1947. Suggested Retail $0.25. Kitchen Table, 1-13/16" x 2-3/8" x 3-5/8", (No. 67), 1945-1953 and 1956. Suggested Retail $0.10. Chair, 2-5/8" x 1-1/4" x 1-3/8", (No. 63), 1945-1947. Suggested Retail $0.05. White and black, Renwal Mfg. Co., Inc., USA.

Stove, 2-5/8" x 2-7/8" x 2", (No. 69), 1948-1953 and 1956. Suggested Retail $0.29. Refrigerator, 3-1/2" x 1-15/16" x 1-5/16", (No. 66), 1948-1953 and 1956. Suggested Retail $0.29. Kitchen Sink, 2-11/16" x 3" x 2", (No. 68). Suggested Retail $0.29. Table, 1-13/16" x 2-3/8" x 3-5/8", (No. 67), 1945-1953 and 1956. Suggested Retail $0.10. Chair, 2-5/8" x 1-1/4" x 1-3/8", (No. 63), 1948-1956, white and red, Renwal Mfg. Co., Inc., USA.

Deluxe Sink (No. 22), 6" x 5 1/2" x 4 1/8", Deluxe Refrigerator (No. 24), 7" x 4 1/4" x 2 1/2", Deluxe Stove (No. 26), 5 7/8" x 5" x 3 7/8", Renwal Mfg. Co., Inc., USA, 1948-1950. Suggested retail $1.89 each.

Deluxe Sink (No. 22), 6" x 5 1/2" x 4 1/8", with water tank and working faucets, includes painted wood with paper label, bar of soap, can of soap powder, and box of cleanser, Deluxe Refrigerator (No. 24), 7" x 4 1/4" x 2 1/2"), with two removable shelves and a removable two piece ice tray, includes painted wood with paper label; package of butter, cheese and eggs and a bottle of milk, Deluxe Stove (No. 26), 5 7/8" x 5" x 3 7/8", with hide-away lid that covers burners, includes three assorted pots and pans, all three Deluxe kitchen pieces white with black trim, Renwal Mfg. Co., Inc., USA, 1948-1950. Suggested Retail $1.89 each.

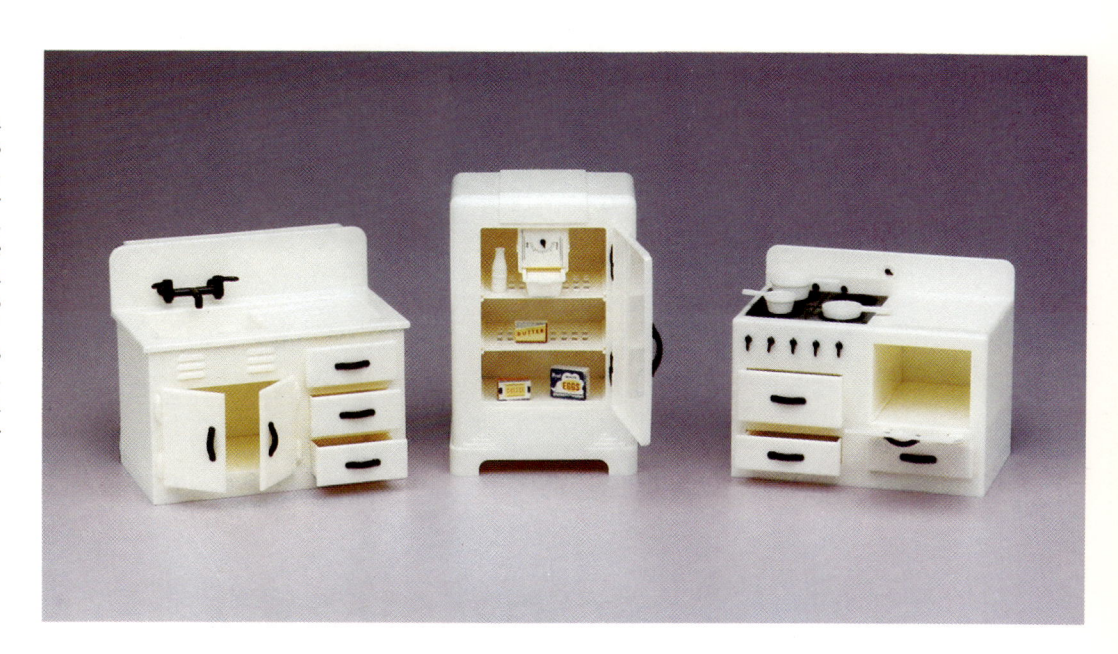

Size Comparison, Deluxe Stove (No. 26), 5-7/8" x 5" x 3-7/8", 1948-1950, and Standard Stove (No. 69), 2-5/8" x 2-7/8" x 2", 1945-1947, Renwal Mfg. Co., Inc., USA.

DOLL HOUSE FURNITURE 257

Barrel Chair, 2-3/16" x 2-3/8" x 2-5/16", (No. 77), 1945-1953 and 1956. Suggested Retail $0.19. Sofa, 2-1/4" x 5-1/16" x 2-3/16", (No. 78), 1945-1953 and 1956. Suggested Retail $0.25. Cocktail Table, 1-1/4" x 1-7/16" x 2-5/16", (No. 72), 1945-1953 and 1956. Suggested Retail $0.10. Floor Lamp 3" x 1-1/4", (No. 70), 1945-1956. Suggested Retail $0.10. Club Chair, 2-9/16" x 2-5/8" x 2-1/2", (No. 76), 1945-1953 and 1956, assorted colors and color combinations, Renwal Mfg. Co., Inc., USA.

Console Radio, 2-1/4" x 1-9/16" x 7/8", (No. 79), 1945-1947. Suggested Retail $0.10. Fireplace, 3" x 4" x 1-1/2", (No. 80), 1945-1953. Suggested Retail $0.39. Radio Phonograph, 2-1/4" x 1-5/8" x 1", 1948-1953, mahogany color, Renwal Mfg. Co., Inc., USA.

Piano and Bench, Piano 2-1/4" x 3-1/8" x 3-11/16", mahogany color, (No. 74), 1945-1951, Bench, 1-1/4" x 7/8" x 2", mahogany color, (No. 75), 1945-1953. Suggested Retail $0.05. Table Lamp, 1-3/8" x 3/4", assorted colors, (No. 71), 1945-1956. Suggested Retail $0.05. End table, 1-11/16" x 1-3/8", mahogany color, (No. 73), 1945-1956. Suggested Retail $0.10. Renwal Mfg. Co., Inc., USA.

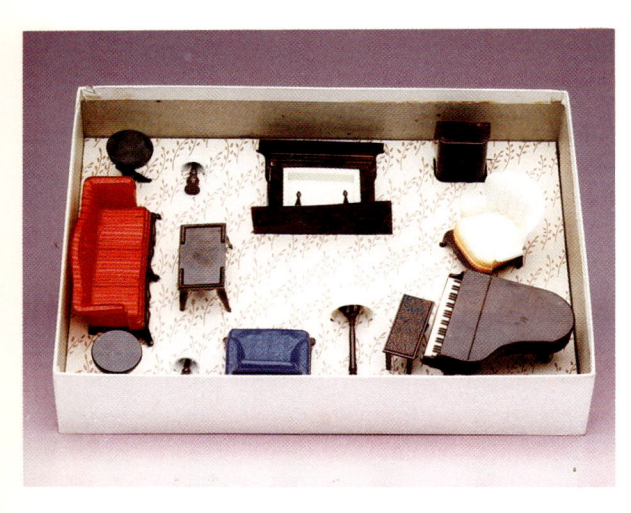

Living Room Set, box 14-1/4" x 10-1/4" x 3", includes sofa, two end tables, two table lamps, cocktail table, club chair, fireplace, floor lamp, console radio, barrel chair, piano bench, and lithographed fold out cardboard living room, assorted colors and color combinations, (No. unknown), Renwal Mfg. Co., Inc., USA.

Assorted Doll House Sets, boxes 12-1/4" x 8-3/4" x 2-1/2" and 14-1/4" x 10-1/4" x 3", Renwal Mfg. Co., Inc., USA, 1945-1947.

Rocking Chair, 2-5/8" x 2-1/4" x 1-3/4", (No. 65), 1949-1955. Suggested Retail $0.10. Baby Bath, 3-1/4" x 2-1/8" x 2-5/8", (No. 122), 1947-1955. Suggested Retail $0.25. Highboy, 3-3/16" x 2-5/16" x 1-1/16", (No. 85), 1945-1948, assorted colors and color combinations, Renwal Mfg. Co., Inc., USA.

Cradle, 2-1/16" x 2" x 2-7/8", (No. 120) with doll insert, 1947-1950, (No. 119) with spread insert, 1947-1953 and 1956. Suggested Retail $0.19. High Chair, 3-5/8" x 2-1/16" x 2-1/16", (No. 30), 1947-1952; Play Pen, 3-1/4" x 3-1/4" x 1-3/ , (No. 118), 1947-1953 and 1956. Suggested Retail $0.25. Assorted colors and color combinations, Renwal Mfg. Co., Inc., USA.

Stroller, 3-1/2" x 2-1/2" x 1-3/4", (No. 87), 1949-1953, Carriage, 2-7/8" x 1-3/4" x 3-9/16", (No. 115) with doll insert, 1947-1950, (No. 114) with spread insert, 1947-1953 and 1956, assorted color combinations, Renwal Mfg. Co., Inc., USA.

DOLL HOUSE FURNITURE

Tricycle, 2-1/16" x 2-9/16" x 2", (No. 7), 1948-1953; Swing, 4-1/2" x 3" x 3", (No. 19), 1948-1951; See-Saw, 4-3/8" x 1-3/8" x 1", (No. 21), 1948-1952. Suggested Retail $0.10. Slide, 6" x 3-1/2" x 1-1/4", (No. 20), 1948-1952. Suggested Retail $0.19. Kiddie Car, 2-5/8" x 2" x 1-3/4", (No. 27), 1949-1951, assorted color combinations, Renwal Mfg. Co., Inc., USA.

Student's Desk, 2-3/8" x 2-5/8" x 1-1/2", (No. 33), 1947-1950; Teacher's Desk, 4" x 2-1/2" x 2-1/4", (No. 34), 1947-1950; Teacher's Chair, 3" x 2-3/8" x 2", (No. 35), 1947-1950, assorted colors, Renwal Mfg. Co., Inc., USA.

Sewing Machine, 3-3/8" x 2" x 1-3/4", drawers open, head drops down, treadle works, needle goes up and down, assorted color combinations, Renwal Mfg. Co., Inc., USA (No. 89), 1949-1955. Suggested Retail $0.29.

Examples of Renwal's Decorated Furniture, Kitchen Table, (No. 67). Suggested Retail $0.15. Kitchen Sink, (No. 68). Suggested Retail $0.39. Sofa, (No. 78). Suggested Retail $0.29. Bed, (No. 81). Suggested Retail $0.29. Club Chair, (No. 76). Suggested Retail $0.25. Assorted colors and color combinations with hand painted designs, Renwal Mfg. Co., Inc., USA, 1954-1955.

Ironing Board and Iron, 3-1/2" x 1-3/4" x 1-1/4", folds flat, assorted colors, Renwal Mfg. Co., Inc., USA (No. 32), 1949-1953. Suggested Retail $0.15.

Washing Machine, 3-1/2" x 2-1/2", wringer and agitator turn, assorted color combinations, Renwal Mfg. Co., Inc., USA (No. 31), 1949-1955. Suggested Retail $0.29.

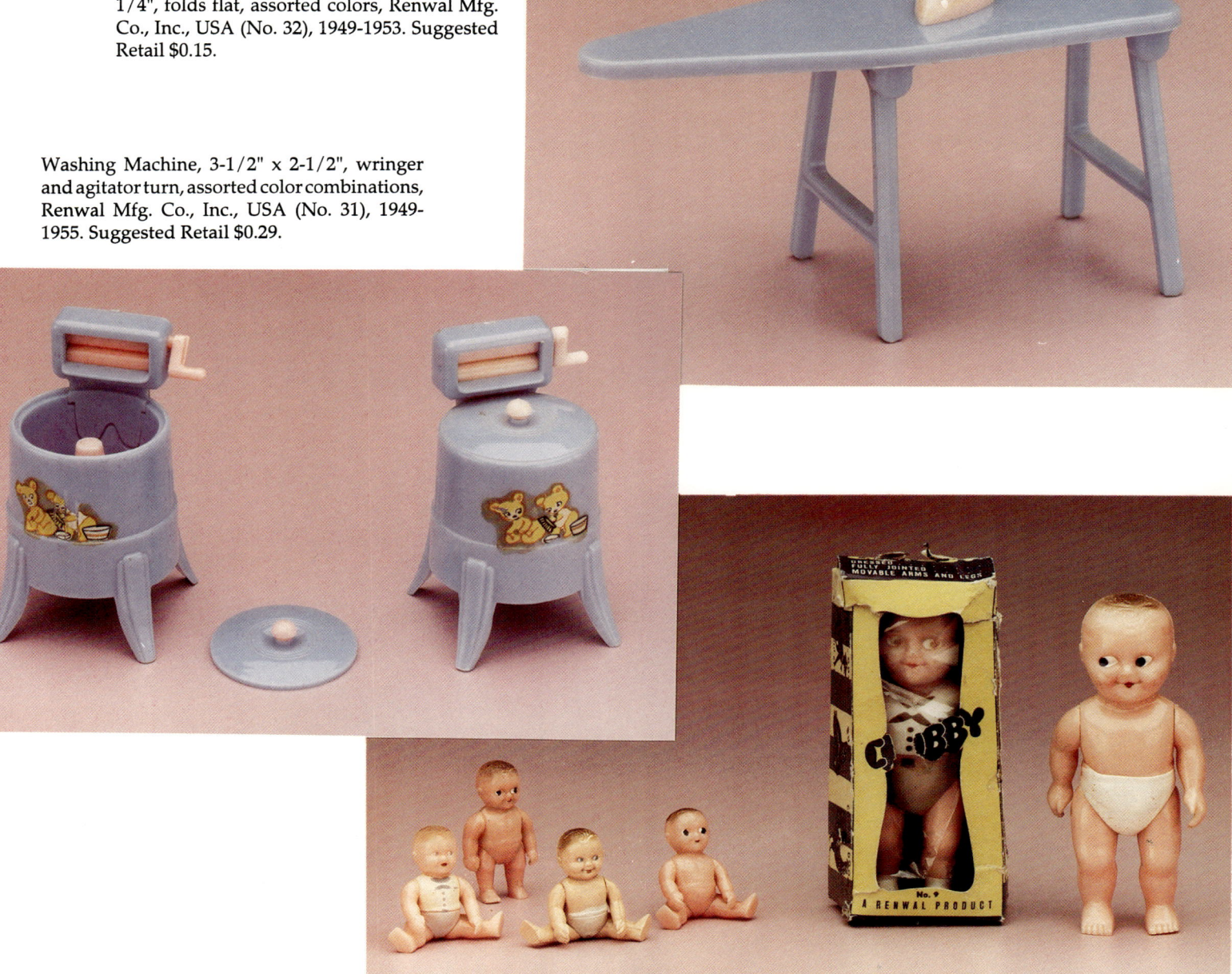

Doll, 2-1/4" tall, with painted clothing (No. 8), 1947-1956. Suggested Retail $0.10. Doll, 2-1/4" tall, with painted diapers (No. unknown), 1947-1949. Suggested Retail $0.10. Doll, 2-1/4" tall, with naked body (No. 5), 1950-1954. Suggested Retail $0.10. Doll (Chubby), 5-1/8" tall, with painted clothing (No. 9), 1948-1951. All of the 2-1/4" dolls have No. 5 on their backs, Renwal Mfg. Co., Inc., USA.

Doll Family (unpainted), box 11-1/2" x 6" x 1", blue brother missing, assorted colors, Renwal Mfg. Co., Inc., USA (No. 544), 1954-1955. Suggested Retail $1.29.

DOLL HOUSE FURNITURE 261

Mother Doll, 4-1/8" tall, painted, (No. 43), 1949-1952. Brother Doll, 3-3/4" tall, painted, (No. 42), 1949-1952. Doctor Doll, 4-1/4" tall, painted, (No. 441), 1951-1952. Nurse Doll, 4-3/8" tall, painted, (No. 431), 1951-1952. Father Doll, 4-1/4" tall, painted, (No. 44), 1949 - 1952. Sister Doll, 3-5/8" tall, painted, (No. 41), 1949-1952. Not shown, Policeman Doll, 4-1/2" tall, painted, (No. 442), 1951-1952, and Mechanic Doll, 3-7/8" tall, painted, (No. 421), 1951-1954, Renwal Mfg. Co., Inc., USA. Suggested Retail $0.29 each.

Kitchen Accessories, 1" to 2-1/2" long, coffee pot missing lid, butter dish, pressure cooker, ice tray, covered dish, skillet, drainer, red, manufacturer unknown, USA, late 1940s to early 1950s.

Carpet Sweeper, 6" x 2-3/8" x 1-1/2", accurate working model of a real carpet sweeper with bristle brush that revolves when sweeper is pushed and two opening doors on the bottom to remove dust, assorted color combinations, Teeny Toy Co., USA, late 1940s to early 1950s.

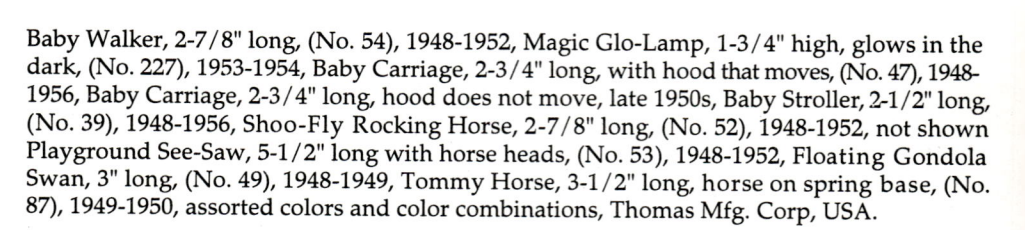

Baby Walker, 2-7/8" long, (No. 54), 1948-1952, Magic Glo-Lamp, 1-3/4" high, glows in the dark, (No. 227), 1953-1954, Baby Carriage, 2-3/4" long, with hood that moves, (No. 47), 1948-1956, Baby Carriage, 2-3/4" long, hood does not move, late 1950s, Baby Stroller, 2-1/2" long, (No. 39), 1948-1956, Shoo-Fly Rocking Horse, 2-7/8" long, (No. 52), 1948-1952, not shown Playground See-Saw, 5-1/2" long with horse heads, (No. 53), 1948-1952, Floating Gondola Swan, 3" long, (No. 49), 1948-1949, Tommy Horse, 3-1/2" long, horse on spring base, (No. 87), 1949-1950, assorted colors and color combinations, Thomas Mfg. Corp, USA.

Playground Triple Swing, 7" long, 1948-1950, uncataloged; Playground Slide, 4-3/4" long, (No. 55), 1949-1950; Hammock, 4-1/2" long, (No. 70), 1949-1950; Playground Single Swing, 4-1/2" long, (No. 51), 1948-1953, not shown; Playground See-Saw Swing, 4-3/4" long, (No. 69), 1949-1950; Ferris Wheel, 6-1/8" tall with four seats, (No. 85), 1950-1952; Playground Duo-Swing, 5-1/2" long, (No. 64), 1949-1950, assorted colors and color combinatins, Thomas Mfg. Corp., USA.

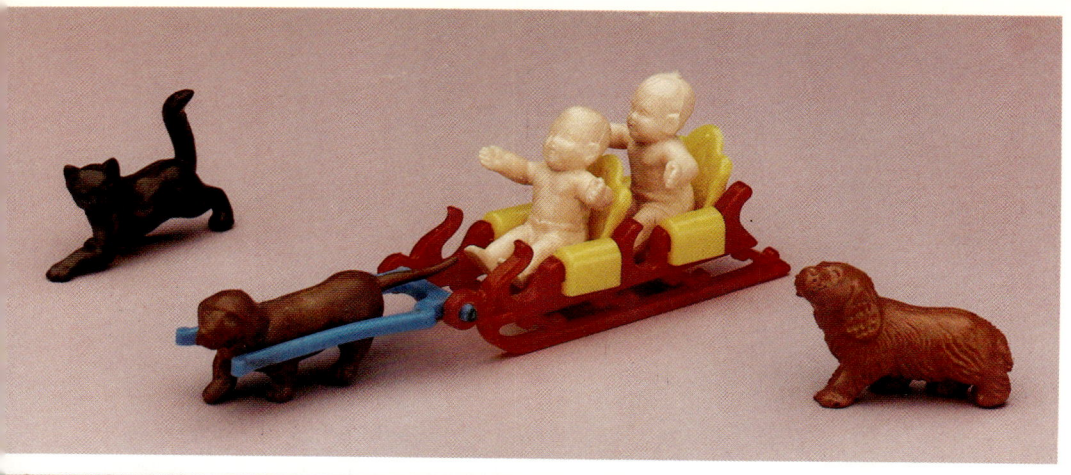

Dog Sled and Two Boy Dolls, 5" long, comes with one of two different vinyl dogs and two 1-1/2" tall vinyl boy dolls (wrong dolls shown in photo), assorted color combinations, (No. 83), 1949-1950; Flexible Dog and Cat Assortment, two different dogs and one cat, six dozen to a carton, brown and tan, (No. 50), 1949-1956, Thomas Mfg. Corp., USA.

Scooter Wagon, 3" x 1-3/4" x 1", assorted colors, Thomas Mfg. Corp., USA (No. 37), 1947-1950.

Doll House Playtime Family, card size, 7" x 7", includes vinyl mama doll, 3-1/4" tall; vinyl papa doll, 3-3/8" tall; vinyl baby doll 1-3/8" tall; vinyl boy or girl doll, 1-1/2" tall; vinyl painted boy or girl doll, 2" tall and one of either two dogs or a cat. Thomas Mfg. Corp., USA (No. 146), 1953-1956, all figures were introduced separately by 1950.

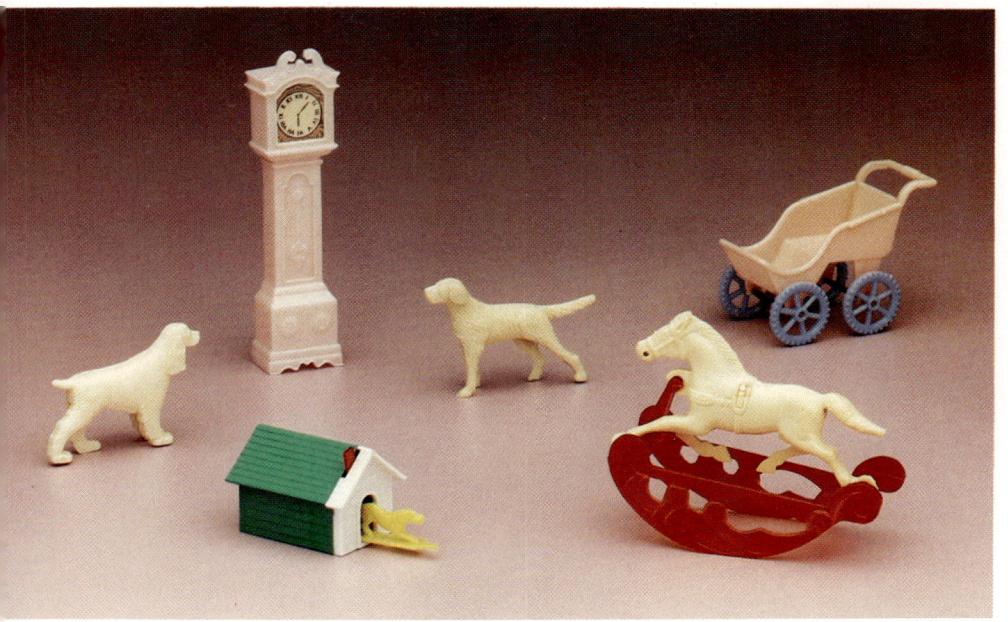

Miscellaneous Pieces; Dogs, 2-1/4" long, Caldwell Products, USA, late 1940s to early 1950s; Grandfathers Clock, 4-3/4" tall, Sun Dial Products, USA, 1948 to early 1950s; Dog in Dog House, 2-3/4" long, push chimney and dog jumps out, manufacturer unknown, USA, early 1950s; Rocking Horse, 3-3/4" long, Hardy Plastics and Chemical Corp., USA, early 1950s; Baby Stroller, 3" x 1-5/8" x 2-3/8", Allied Molding Corp., USA, 1949 to early 1950s.

DOLL HOUSE FURNITURE 263

CHAPTER 16
POTPOURRI

The use of plastics was certainly not restricted to the major toy categories already discussed in this book. There were hundreds of other successful applications, many of which are worthy of being collected and displayed today.

For those of us growing up in the late forties and early fifties, playing cowboys and Indians wasn't a passing fad, it was a national pastime. The adventures of Hopalong Cassidy, Gene Autry, The Lone Ranger and Roy Rogers, seen weekly on television by millions of would-be cowboys and cowgirls, inspired hundreds of toys, many of them plastic.

Plastic toys bearing the name Roy Rogers were far and away the most popular. Of the hundreds of toys endorsed by the "King of the Cowboys" some of the most imaginative and successful were produced by the Ideal Toy Corporation between 1955 and 1959.

Starting with the "Roy Rogers Fix-It Stage Coach" which debuted in 1955, Ideal quickly expanded its Roy Rogers line by offering a chuck wagon and jeep, a horse trailer and jeep, a buckboard, a western communicating telephone set and a western dinner set. All of these are highly collectible and eagerly sought after by both plastic collectors and collectors of Roy Rogers memorabilia.

Roy Rogers and Dale Evans with Benjamin F. Michtom, Chairman of the board, Ideal Toy Corp., USA, 1956.

Another interesting group of toys produced by Ideal featured a hand operated talking mechanism, licensed from its inventor Ted Duncan. The first talking toys to feature this miniature record player were introduced in 1954. The talking toys included a police car, a wall phone, a train and a remote controlled robot named Robert. When the crank on Robert's back was turned, he repeated "I'm Robert the Robot, the Mechanical Man. Drive me and steer me wherever you can." "Robert the Robot" was one of Ideal's all-time biggest hits and remained in the line for an amazing six years.

In 1955, Ideal added seven more talking toys to their plastics line. These included the "Hickory Dickory Clock" talking toy, the "Three Blind Mice" talking toy, a "Dragnet" talking police car based on the popular TV show and four *Little Golden Book* talking toys.

The *Little Golden Book* talking toys were based on the most beloved "Golden Book" characters. The "Poky Little Puppy", the "Saggy, Baggy Elephant", "Scuffy, the Tugboat" and "Tootle, the Train" all repeated lines from their famous books when their hand cranks were turned.

Given the tremendous popularity of the *Golden Books* series at the time, it's hard to believe that more of these delightful toys didn't survive. Perhaps the mechanism broke easily and they were simply tossed out like so many other childhood memories or perhaps they were never accepted by the toy buying public.

In either case, they were only offered for two years, and those that remain are considered real treasures by both plastic and *Golden Book* collectors.

The subject of plastic figures made from either vinyl or polyethylene has purposely been omitted from this book.

The wonderful playsets offered by companies like Marx and Ideal during the 1950s contained a wide variety of figures made from unbreakable vinyl and polyethylene plastics. They provided millions of children with the means to recreate famous battles of history or to reenact the latest episode of their favorite TV show. These figures are highly collectible today and given the number of different playsets and figures produced, certainly are beyond the scope of this book and worthy of their own book.

There is, however, one manufacturer of cellulose acetate figures that should be mentioned in any discussion of the history of plastic toys. The Bergen Toy and Novelty Company, also known as Beton, a shortened version of its full name, was established around 1935 by Charles Marcak. Marcak's original goal was to produce a line of lead soldiers which were the toy soldiers of choice at the time.

Originally located in Carlstadt, New Jersey, Beton introduced the first plastic figures in 1938. The line included soldiers, cadets and cowboys and Indians. With the exception of some of its animals, all Beton figures had bases. The earliest figures had glue-on oval bases, next came glue-on rectangular bases and finally bases that were an integral part of the figure.

Beton itself did not own any injection molding machines before or during the war. All molding appears to have been done by the Columbia Protektosite Company, Inc., also located in Carlstadt.

In an article in the August 1939 issue of *Modern Plastics* magazine, another company is credited with "experimenting with a line of molded plastic soldiers, Indians and such." The company credited is the Universal Plastics Corporation, and the figures shown are similar, if not identical, to Betons No. 515 Rifleman, marching position, No. 513 Infantryman, saluting, No. 515 Infantryman, charging, No. 729 Indian Chief, No. 730 Indian Warrior and No. 731 Indian with Drawn Bow. All figures shown have rectangular glue-on bases and interestingly enough, all of the soldiers have soft caps on instead of helmets.

Other figures shown in the same article are credited as being molded by the Columbia Protektosite Company, Inc. These figures include Beton marching cadets and No. M412 Cowboy Mounted on Bronco and No. M413 Road Agent Mounted on Bronco.

Perhaps Beton used both of the above custom molders or perhaps Universal was developing its own line of figures and Beton purchased the molds and added them to its own. These are the kinds of sightings that can keep toy historians up all night until the correct sequence of events is verified.

Beton's first figures were molded from Tenite, a cellulose acetate from the Tennessee Eastman Corporation. They offered features their lead counterparts could not lay claim to. Light in weight and practically unbreakable, they were also non-toxic should little Johnny decide to have one for lunch.

While these were definite pluses, there were also minuses as their size and limited number of poses did not seem to fit in with the popular Manoil and Barclay lead figures which accounted for most of the toy armies of the time.

1951 Bergen Toy and Novety Co. Catalog.

Milk Wagon and Horse, 7-3/4" x 2-1/2" x 2-7/8", with two miniature milk crates, white, red, yellow and gray, plastic wheels, All Metal Products Co., USA (No. 4002), early 1950s. Suggested Retail $0.59.

Offered in both painted and unpainted versions, the painted variety used only a dash of paint here and there, offering further negative comparison.

America's entry into the war would eliminate all metal soldier production by April, 1942 and, except for several companies producing composition soldiers, Beton was without competition through the fall of 1945.

With a seemingly unlimited supply of scrap material available, Beton filled the shelves of dime stores across America during the war, limited only by production capacities.

After the war, Beton began molding its own figures in its new Hackettstown, New Jersey plant. It continued to be successful into the mid 1950s and some of its best figures like the farm and circus series were introduced during this post-war period.

The toy figure market was going to be huge and the number one and two toy companies wanted their share. Marx and Ideal both entered the figure market in the early 1950s. Their figures cost less, were truly unbreakable and offered a much broader choice of themes and poses. Post-war consumers demanded more realism and the latest in military hardware. Ideal and Marx delivered.

Beton, who continued to market many pre-war designs and who updated many of their military figures by simply changing the helmet style, began to lose sales. When Japan started copying some of Beton's figures in the late 1950s, the company was sold to Rel Plastics of East Paterson, New Jersey. Beton figures are very collectible today and some of their early post-war packaging ranks near the top for artistic design and sales appeal.

Other successful toy applications included a wide variety of character items, toy guns, doctor and dentist sets, garden sets, tea sets and miniature working appliances.

From ambulances to zebras, toys from the "hard plastic" era offer something for every collector. Their cheerful colors and intricate designs should be a welcome addition to any collection, and a pleasant reminder of lazy summer afternoons spent in one's favorite dimestore.

Liftbridge 'N Boat, 9-1/2" x 5-1/2" x 1-1/2", when crank is turned boat goes around and bridge raises and lowers, assorted color combinations, Amerline, USA (No. 568), 1953 to mid 1950s. Suggested Retail $0.98.

Wagon and Horse, 7-3/4" x 3" x 2", wagon body made from dump truck bed, assorted colors, plastic wheels; Coal Wagon and Horse, 7-3/4" x 3" x 2-5/8", wagon body made from coal truck bed, assorted colors, plastic wheels, Banner Plastics Corp., USA. Early 1950s. Suggested Retail $0.29 each.

Circus Wagons, 2-1/4" x 1-1/8" x 1-1/2", assorted colors, stationary wheels, Banner Plastics Corp., USA, 1946 to early 1950s.

Assorted Beton Figures, 2-3/4" tall, painted details, Bergen Toy and Novelty Co., USA, 1940s to 1950s.

Railroad Folks, standing figures 2-3/4" tall, painted details, bulk packed two dozen assorted pieces per 9" x 4" x 3" box, Bergen Toy and Novelty Co., USA (No. 584), early to mid 1950s.

POTPOURRI 267

Assorted Animals; Moose 3-1/2" tall; Hippopotamus 3-3/4" long; painted details, bulk packed one dozen pieces of one animal per 9" x 4" x 3" box, Bergen Toy and Novelty Co., USA, early to mid 1950s.

Dairy Set, box 11-1/2" x 8-1/2" x 2-1/4", Farmer, 3" tall; Bull, 4" long, painted details, Bergen Toy and Novelty Co., USA (No. 171), early to mid 1950s.

Small Circus Animals, box 11-1/2" x 8-1/2" x 2-1/4"; Lion Tamer, 2-1/2" tall; Alligator, 5-1/2" long, painted details, Bergen Toy and Novelty Co., USA (No. Z-250-S), early to mid 1950s.

Infantry Footmen, box 12" x 5" x 4-1/2", average figure 2-3/4" tall, painted details, Bergen Toy and Novelty Co., USA (No. 500), early to mid 1950s.

Cowboys and Indians, box 16" x 11-1/2" x 2", standing figures 2-3/4" tall, painted details, Bergen Toy and Novelty Co., USA (No. M-383), early to mid 1950s.

Infantry and Cavalry, box 16" x 11-1/2" x 2", standing figures 2-3/4" tall, painted details, Bergen Toy and Novelty Co., USA (No. M-384), early to mid 1950s.

Stagecoach, 9-1/2" x 2-1/2" x 3-5/8", with driver, assorted colors, plastic wheels, Hardy Plastics and Chemical Corp., USA, 1950 to mid 1950s.

Climbing Monkey, 6-3/4" tall, stretch the string tight and watch Jocko scramble up and down, assorted colors, Ideal Novelty and Toy Co., USA (No. MON-69), 1945-1947. Suggested Retail $0.69.

POTPOURRI 269

Six Shooter, 8-1/4" long, with clicking trigger and revolving cylinder, black, Ideal Novelty and Toy Co., USA, (No. GU-100), 1945-1947. Suggested Retail $1.00.

Chirping Chick, 3" tall, when egg is squeezed, chirping chick pops out, (No. EC-40), 1949-1950. Suggested Retail $0.39. Pecking Chick in Cart, 5-3/4" x 4-3/4" x 3", when cart is pulled, chick's legs scratch and beak pecks in and out of tray, (No. 3457), 1949-1955. Suggested Retail $0.39. Hen in Basket, 4-1/2" x 4-1/2" x 3", slot for Easter ribbon or use as bank, (No. 3440), 1953-1955, assorted color combinations, Ideal Novelty and Toy Co., USA.

Giant Piggy Bank, 12" tall, with coin slot in back and removable coin trap in bottom, pink, painted details, Ideal Novelty and Toy Co., USA (No. PIB-200), 1950, (No. 4170), 1951-1957. Suggested Retail $1.99.

Suzie Plastic Doll, 5-1/2" tall, a rattle toy, assorted colors with painted details, (No. PD-40), 1947-1948. Suggested Retail $0.39. Junior Shaver, 4" x 2", when wound up, shaver simulates a real electric razor sound and action, assorted color combinations, (No. ES-70), 1948-1950, (No. 3875), 1951-1958. Suggested Retail $0.69. Telephone with Pad and Pencil, 7-1/2" x 3-1/2", assorted color combinations, (No. 4261), 1952-1953. Suggested Retail $0.79. Standing Plastic Dog, 5-1/2" tall, a rattle toy, assorted colors with painted details, (No. PD-30), 1947-1948. Suggested Retail $0.29. "Suck-A-Thumb", 5-1/2" tall, a rattle toy, assorted colors with painted details, (No. PGS-35), 1947-1948. Suggested Retail $0.35. Ideal Toy and Novelty Co., USA.

270 PLASTIC TOYS

Hot Dog Wagon, 9" x 3" x 6", with sliding doors, opening food compartment lids, removable ice cream compartment lid, detachable umbrella, assorted color combinations, plastic wheels, Ideal Novelty and Toy Co., USA (No. FW-80), 1949-1950. Suggested Retail $0.79.

Television Bank, 5" x 3" x 7", with antenna that can be raised and lowered, removable coin trap on bottom, assorted colors, Ideal Toy Corporation, USA, (No. 4159), 1953, antenna not shown. Suggested Retail $0.80.

Pepsi Cola Wagon, 7-1/2" x 6-1/2" x 3", with sliding doors, opening food compartments, lids, removable ice cream compartment lid, detachable umbrella and 6" painted plastic man, white with red umbrella lids and wheels, Pepsi Cola decal, Ideal Novelty and Toy Co., USA (No. PW-100), 1950, (No. 4243), 1951.

Hopalong Cassidy and Topper, 4-3/4" x 5-5/8", removable Hoppy with movable arm and removable hat, Topper has bead chain bridle, white horse with painted black saddle, Hoppy painted in authentic colors, Ideal Novelty and Toy Co., USA (No. HC-100), 1950, (No. 3150), 1951-1957. Suggested Retail $1.00.

Howdy Doody Sand Forms, each of four molds approximately 5" high, with 8-1/4" long shovel, assorted colors, Ideal Toy Corporation, USA, (No. 4219), 1953-1955. Suggested Retail $0.60.

Robert, the Robot, 14" x 6" x 6-1/4"; remote-controlled robot moves forward or backward when crank is turned, and left or right when trigger is pulled; arms go up and down and hands open and close manually; off/on switch on chest controls battery powered light for eyes and antenna; opening tool chest on front of robot below light switch contains die cast hammer, screwdriver and wrench; when crank is turned on Robot's back he says, "I am Robert, the Robot, the Mechanical Man. Drive me and steer me wherever you can." Silvertone finish with red arms, control box and other details, concealed rubber wheels, Ideal Toy Corporation, USA, (No. 4049), 1954-1959. 1955 tool box and tools removed, 1956 clear plastic antenna removed. Suggested Retail $6.00.

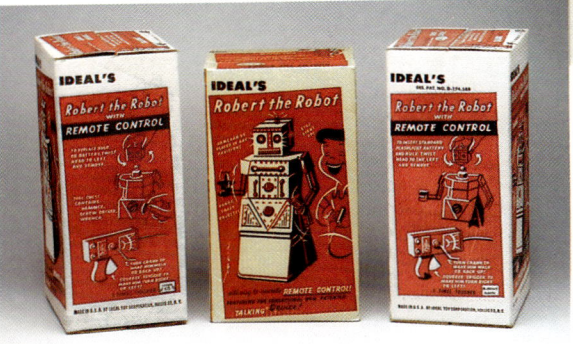

Robert, the Robot, box variations, L to R, first, third and second, Ideal Toy Corp., USA (No. 4049), 1954-1959.

Talking FBI Car, 14-1/2" x 5-1/2" x 5-3/4", with removable dome that has a battery powered search light on top which can also be removed and used as flash light. Detailed interior includes two front seats, steering wheel, removable phone that hangs from dash, opening storage compartment that holds two pistols, flash camera, tommy gun, riot gun, binoculars and two rifles, swivel chair and map table in rear. When crank on trunk is turned car says, "Calling all cars. Go to 1234 Main Street. That is all." White body with metallic blue paint, FBI decal on one door, plastic wheels, Ideal Toy Corporation, USA, (No. 3072), 1954. Suggested Retail $4.00.

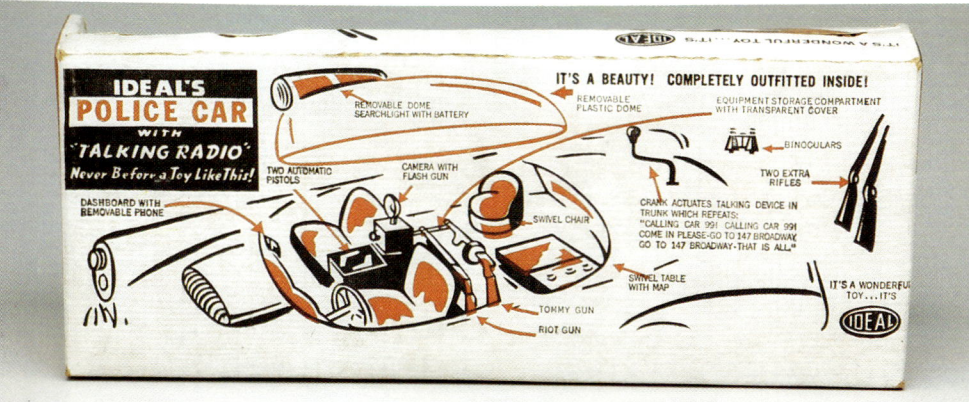

Talking FBI car, box detail, Ideal Toy Corp., USA (No. 3072), 1954.

272 PLASTIC TOYS

"Dragnet" Talking Police Car, 14-1/2" x 5-1/2" x 5-3/4", with removable dome which has a battery powered search light on top that can also be removed and used as a flash light, detailed interior includes; two front seats, steering wheel, removable phone that hangs from dash, opening storage compartment that holds; two pistols, flash camera, tommy gun, riot gun, binoculars, and two rifles, swivel chair and map table in rear, when crank on trunk is turned, car says, "Calling Car 99, calling Car 99, come in please. Go to 147 Broadway. That is all." Combinations of black body and painted white doors or painted white front and rear of car, chrome finished grill and headlights, "Dragnet" decal on one side, plastic wheels, Ideal Toy Corporation, USA, (No. 3006), 1955-1957. Suggested Retail $4.00

Hickory Dickory Clock Talking Toy, 10" x 8-1/2", as a hand-crank is rotated, a mouse scampers around the face of the clock until the record reciting "Hickory Dickory Dock" stops, hands may be moved to teach telling of time, assorted color combinations, Ideal Toy Corporation, USA, (No. 4300), 1955-1956. Suggested Retail $3.00.

Talking Train, 12-1/2" x 9-1/4" x 2", when crank is turned, plastic train goes around track on lithographed tin base and train station says, "All Aboard" with sounds of trains and whistles in background, Ideal Toy Corporation, USA, (No. 4273), 1954-1956. Suggested Retail $2.00.

Talking Wall Phone, 9-5/8" x 6-3/8" x 4", when crank on side is turned, phone says, "Operator. Operator. Number please. Number please. What number are you calling? Thank you. I am ringing your number." Deposited coins automatically fall through return slot when receiver is off the hook, black, Ideal Toy Corporation, USA, (No. 4263), 1954-1957. Suggested Retail $4.00.

"Saggy, Baggy, Elephant", 8" high, when hand-crank is rotated, "Saggy" recites from the Golden Book story, authentic "Saggy" colors, Ideal Toy Corporation, USA, (No. 4305), 1955-1956. Suggested Retail $1.60.

Three Blind Mice Talking Toy, 12-1/2" x 9-1/4", as the hand-crank is rotated, the "farmer's wife" pursues three mice who go round and round as the record recites "Three Blind Mice", lithographed tin and assorted color combinations, Ideal Toy Corporation, USA, (No. 4301), 1955-1956. Suggested Retail $3.00.

"Poky Little Puppy", 7-1/2" high, when hand-crank is rotated, "Poky" recites from the Golden Book story, authentic "Poky" colors, Ideal Toy Corporation, USA, (No. 4304), 1955-1956. Suggested Retail $1.60.

"Scuffy, the Tugboat", 8" high, when hand-crank is rotated, "Scuffy" recites from the Golden Book story, authentic "Scuffy" colors, Ideal Toy Corporation, USA, (No. 4302), 1955-1956. Suggested Retail $1.80. Not shown, "Tootle, the Train", (No. 4303), 1955-1956. Suggested Retail $1.80.

Sir Galahad, knight 9-1/4" tall, horse 11" tall, removable knight in simulated armor has removable helmet with adjustable visor, red battle axe, long-blade sword and spurs, horse is black and protected by five removable pieces including saddle of red armor, black vinyl removable harness, bridle, stirrups, spurs and belt for sword, Ideal Toy Corporation, USA, (No. 3149), 1955. Re-numbered (No. 3146), 1956-1957. Suggested Retail $6.00.

FIX-IT Stage Coach, 15" x 4-1/4" x 6-1/4", vinyl driver with beard, two horses, opening doors, and the following ; two spare wheels, jack, mallet, two wheel wedges, pry bar, wrench, to change wheels, removable vinyl harnesses, reins and whip, rifle, two piece strong box and two piece trunk, a yellow vinyl sheet covers rear compartment, all wheels may be changed, red and black with painted gold trim, gray horses, plastic wheels, Ideal Toy Corporation, USA, (No. 4801), 1953-1954. Also issued as Davy Crockett Alamo Express, (No. 4552), 1955. Suggested Retail $4.00.

Davy Crockett, 4-3/4" x 5-3/8", removable Davy with moveable arm and removable "coonskin" cap, palamino stallion has bead chain bridle, tan horse with painted brown saddle, Davy painted in authentic colors, Ideal Toy Corporation, USA, (No. 3154), 1955-1956. Suggested Retail $1.00.

Davy Crockett Flintlock Pistol, 7-1/2" x 3-1/2", with loud-sounding clicking trigger and hammer, deep mahogany finish with paper Davy Crockett sticker on one side, Ideal Toy Corporation, USA, (No. 4292), 1955. Also issued as Captain Kidd's Silver Pirate Pistol, in brilliant silver finish, (No. 4288), 1954-1955. Suggested Retail $0.20.

POTPOURRI 275

Roy Rogers Fix-It Stage Coach, 15" x 4-1/4" x 6-1/4", with brown bearded driver, two horses, opening doors and the following; two spare wheels, jack, mallet, two wheel wedges, pry bar, wrench to change wheels, removable vinyl harnesses, reins and whip, rifle, two piece strong box and two piece trunk, a yellow vinyl sheet covers rear compartment, red and black with painted gold trim, gray horses, Roy Rogers sticker on one door, Ideal Toy Corporation, USA, (No. 4551), 1955. 1956-1957, stage changed to brown (as shown), and the two spare wheels and vinyl sheet covering rear compartment were dropped. 1958-1960, tan Roy Rogers figure replaced bearded driver. Suggested Retail $4.00.

Roy Rogers Fix-It Chuck Wagon and Nellybelle Jeep, Wagon 19" x 9-3/4", Jeep 7" x 4", with tan vinyl Roy Rogers and Dale Evans, printed vinyl wagon cover that can be removed, wagon wheels that can be removed by means of old-fashioned jack, wrench and pry bar, rear platform drops into place as cook out table, miniature pots and pans included along with two piece chest and strong box, black vinyl reins, two gray horses, brown wagon with green seat, storage box and tailgate, tan plastic wheels, jeep has tan vinyl Pat Brady and "Bullet", hood bobs up and down as it rolls along, yellow with "Nellybelle" sticker on one door, gray plastic wheels, Ideal Toy Corporation, USA, (No. 4553), 1956-1959. Suggested Retail $6.00.

Roy Rogers Horse Trailer and Jeep, Jeep 7" x 4", Trailer 8" long, comical jeep with hood that bounces up and down as it rolls along while Pat Brady drives and Roy sits alongside, trailer may be un-hitched from jeep and has tailgate that drops, Trigger has removable saddle and may be ridden by Roy figure, gray jeep with "Nellybelle" sticker on one door, yellow plastic wheels, red trailer, gray plastic wheels, Ideal Toy Corporation, USA, (No. 4555), 1957-1960. Suggested Retail $4.00.

Roy Rogers Western Electronic Communicating Telephone Set, 9" high, as crank is turned the bell on the other phone rings, carries clear messages from one room to another, uses two flashlight batteries, realistic wood color with black earpieces and mouthpieces, Ideal Toy Corporation, USA, (No. 4269), 1957-1959. Suggested Retail $10.00.

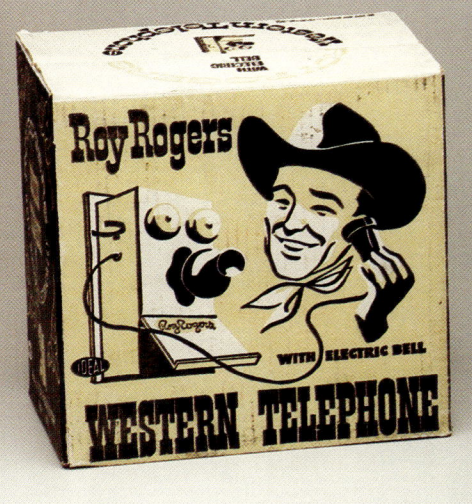

Roy Rogers Western Telephone with Electric Bell, as crank is turned a bell rings electrically, uses one flashlight battery, realistic wood color with black earpiece and mouthpiece, Ideal Toy Corporation, USA, (No. 4257), 1957-1959. Suggested Retail $3.00.

Roy Rogers-Dale Evans Dinner Set, service for two includes plastic wood-grained cups, sugar with cover, creamer, coffee pot with cover and Roy's Double R Brand. Plates and saucers are metal, cooking utensils are black plastic, and knives, forks and spoons are aluminum with simulated wood handles, Ideal Toy Corporation, USA, (No. 4578), 1958-1959. Suggested Retail $3.00. Service for four, (No. 4579), 1958-1959. Suggested Retail $5.00. Service for two, less creamer and cooking utensils, (No. 4577), 1959. Suggested Retail $2.00.

Roy Rogers Buckboard, 16" long, tailgate drops down, with tan vinyl Roy Rogers, Dale Evans and whip, black vinyl reins, wrench and pry bar, mallet, miniature pots and pans, two piece chest and strong box, bucket, lantern and rifle, assorted color combinations, plastic wheels, Ideal Toy Corporation, USA (No. 4550), 1958-1959. Suggested Retail $2.99.

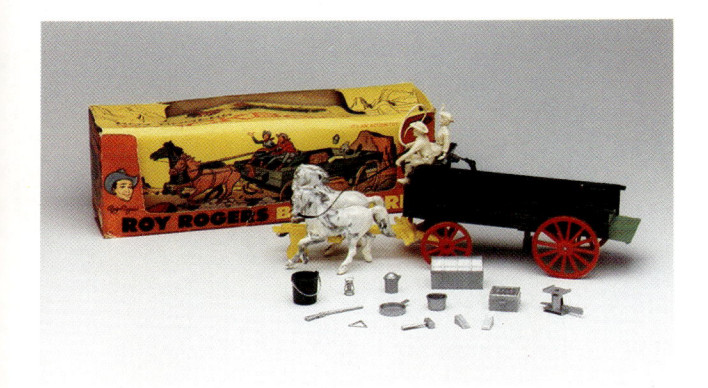

POTPOURRI 277

Juke Box Bank, 4" x 2-1/8" x 6-1/4", to play music, pull out coin receptacle, insert coin and push receptacle in, music box winds in back, assorted color combinations, Ideal Novelty and Toy Co., USA (No. JB-600), 1947-1950, (No. 4229), 1951. Suggested Retail $6.00.

Barky In The Doghouse, 2-3/4" x 2" x 2-1/4", when chimney is pressed, dog barks and moves head side to side, assorted color combinations, Ideal Novelty and Toy Co., USA (No. BD-50), 1948-1949. Suggested Retail $0.49.

Mechanical Washing Machine, 4" x 5" x 8", windup agitator and hand operated mangle, roller and drain pipe, white with red trim, Ideal Novelty and Toy Co., USA (No. 3890), 1948-1950, (No. 3890), 1951-1952. Suggested Retail $2.98.

Garden Set, 9" long, with wheelbarrow, shovel, watering can, rake, hoe and pitchfork, assorted color combinations, Ideal Novelty and Toy Co., USA (No. WB-80), 1949-1950. Suggested Retail $0.49.

278 PLASTIC TOYS

Mechanical Mixer, 9" x 4" x 5", wind-up motor with off/on switch, powerful mixer and juicer with one bowl, assorted color combinations, Ideal Novelty and Toy Co., USA (No. MM-300), 1949-1950, (No. 3848), 1951-1956. Suggested Retail $3.00.

Mechanical Malted Milk Machine, 14" high, windup motor with separate plastic mixing glass, assorted color combinations, Ideal Novelty and Toy Co., USA (No. MA-200), 1950, (No. 3845), 1951-1954, original clear plastic mixing glass not shown. Suggested Retail $2.00.

Spark-Shooting Pirate Gun, when trigger is squeezed, sparks fly from muzzle and hammer is tripped, making loud noise, replaceable flint, assorted color combinations, Ideal Toy Corporation, USA (No. 4286), 1952-1954. Suggested Retail $1.00. Reissued as Davy Crockett Sparking Pistol, "backwoods" colors, (No.4293), 1955.

Color Action TV Blocks, box 12-1/4" x 10-1/2" x 1-5/8", TV, 1-1/2" x 1-1/2" x 1-1/2", when block is moved back and forth figures move, assorted color combinations, Kohner Bros., USA (No. 434), mid 1950s.

Covered Wagon, 7-1/2" x 2-1/2" x 3-3/8", with driver and lithographed metal top, assorted colors, plastic wheels, early to mid 1950s, Lido Toy Co., USA.

POTPOURRI 279

Assorted Animals, 1-5/8" to 3" tall, these whimsical animals were created by Don Manning Studios, assorted colors, Nosco Plastics, USA, 1940 to late 1940s. Similar pieces are still being made today.

King's Knights, box 10-3/4" x 8-3/4" x 1-1/2", mounted knight 3" tall, Lido Toy Co., USA (No. 235), mid 1950s. Suggested Retail $0.59.

Pepsi: Cola Dispenser Bank, 6-3/4" x 3" x 1-7/8", when coin is inserted into slot a miniature 1-1/2" tall bottle of Pepsi is dispensed, comes with three bottles, Louis Marx and Co., USA, early 1950s.

Captain Kidd Pirate Set, box 14-3/8" x 12" x 1-1/2", cutlass 12" long, Lido Toy Co., USA (No. 778), mid 1950s. Suggested Retail $0.98.

280 PLASTIC TOYS

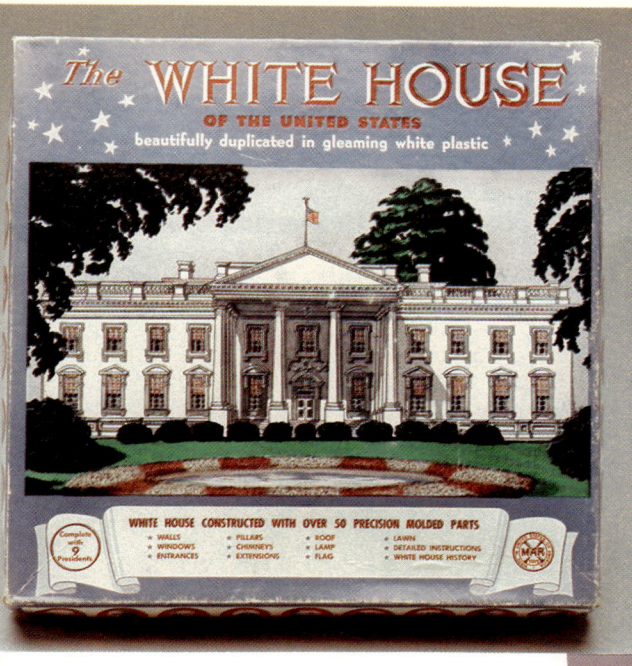

The White House, box 13-3/4" x 13-3/4" x 3-1/2", H.O. scale snap together model of the White House with nine different 2-3/4" tall president figures, white with green lawn area for the front, Louis Marx and Co., Inc., USA, 1953 to mid 1950s. Suggested Retail $3.98.

Covered Wagon, 8" x 3" x 3-3/4", a Marx playset piece, blue, plastic wheels, Louis Marx and Co., Inc., USA, mid 1950s.

Buckboard, 7-3/4" x 3" x 2-1/4", a Marx playset piece, blue, plastic wheels, Louis Marx and Co., Inc., USA, early to mid 1950s.

Davy Crockett, 3-1/2" tall, buckskin color, Plasticraft Mfg. Co., USA, mid 1950s.

POTPOURRI 281

Old Woman in a Shoe, 7-1/4" x 5-1/4" x 6", when pulled, Jack goes up and down in the chimney and children go in and out of the shoe, assorted color combinations, plastic wheels, Renwal Manufacturing Co., Inc. USA (No. 157), 1952-1954. Suggested Retail $1.49.

Play Bank, 3-7/8" x 3-1/2" x 4-1/8", coin is placed in messenger's bag and is deposited in bank when revolving doors are turned, assorted color combinations, Renwal Manufacturnig Co., Inc. USA (No. 158), 1952-1953. Suggested Retail $0.59.

Loco Pete, 9-3/4" x 4-5/8" x 7", when toy is pulled, horse and rider appear to be galloping while rabbit runs alongside, assorted color combinations with painted eyes and teeth, plastic wheels, Renwal Manufacturing Co., Inc. USA (No. 163), 1952-1953. Suggested Retail $1.59.

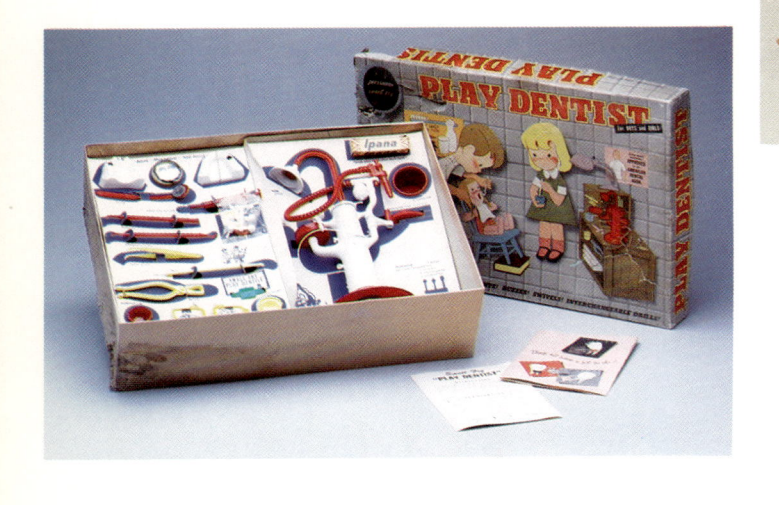

Play Dentist Set, box 16-1/4" x 13" x 5", includes upper and lower dentures, drill and drill bits, novocaine syringe, mirror, numerous instruments, watch, glasses, forceps, filling material, cotton, toothpaste, individual teeth, and working dental unit that lights up and buzzes with real drill sound (Ouch!), assorted colors, Pressman Toy Corp., USA, 1953 to 1954. Suggested Retail $2.99.

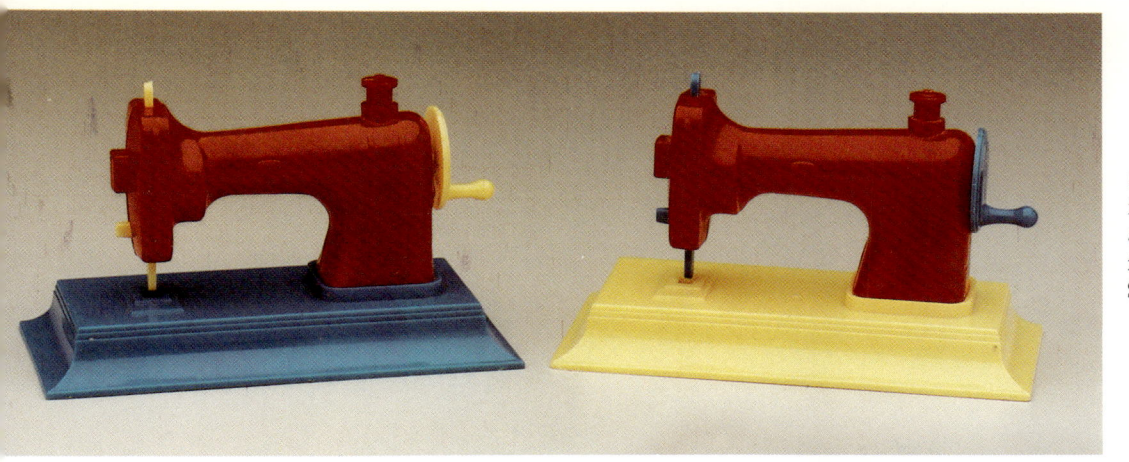

Sewing Machine, 5-3/4" x 3" x 4", when crank is turned, plastic needle goes up and down, assorted color combinations, Renwal Manufacturing Co., Inc. USA (No. 190), 1953-1954. Suggested Retail $0.29.

Renwal Zoo, average size 3-1/2" x 1-1/2" x 3-1/4", four piece pull toy, set includes (No. 180) elephant, (No. 181) camel, (No. 182) horse, and (No. 183) cow, assorted colors, plastic wheels, Renwal Manufacturing Co., Inc. USA (No. 280), 1953, also sold individually.

Cowboy 'n Indian Whistle, 2-5/8" x 3-1/8" x 1", when whistle is blown, cowboy or Indian moves, producing a warbling sound, assorted color combinations, Renwal Manufacturing Co., Inc. USA (No. 451), 1951-1954. Suggested Retail $0.10.

Drawbridge Set, 25-1/4" x 6-3/4" x 4-1/4", with illustrated platform that creates a play environment, including a garage, roads, and a river with a slip for boats, turn knobs to raise and lower bridge, includes two 3-1/8" automobiles, two 4-1/4" boats and two paper flags on wooden poles, assorted colors and color combinations, Renwal Manufacturing Co., Inc. USA (No. 155), 1953-1956. Suggested Retail $2.98.

POTPOURRI 283

U.S. Frontier Set, 26" long, includes limber with opening storage compartments, shell shooting cannon and six cannon balls, seven 2-1/2" to 3-3/8" vinyl cavalry figures and numerous small accessories, gray, yellow, orange and white, Renwal Mfg. Co., USA (No. 296), 1957.

Susy's Superette, box 13-1/2" x 8-1/2" x 6-1/4", includes; check-out stand, cash register, shopping cart, display counter, plastic food and cans, two six packs of soda, parking meter, grocer, housewife and daughter with doll and numerous labels and signs, assorted colors, Kiddie Brush and Toy Co., USA (No. 992), 1956 to late 1950s.

Howdy Doody Figures, 4" tall, includes; Dilly Dally, Howdy Doody, Clarabell the Clown, The Princess and Mr. Bluster. Move lever on the back of the head to make mouth open and close; Clarabell holds a horn you can blow to make a real horn sound. Assorted colors, Tee-Vee Toys, Inc., USA, 1952 to mid 1950s.

Howdy Doody Figures, 4" tall, includes; Dilly Dally, Howdy Doody, Clarabell the Clown, The Princess and Mr. Bluster. Move lever on the back of the head to make mouth open and close; Clarabell holds a horn you can blow to make a real horn sound. Flesh colored plastic with painted details, Flubadub not shown, Tee-Vee Toys, Inc., USA, 1952 to mid 1950s. Suggested Retail $0.98.

Rider and Donkey, 3" x 15/16" x 3-1/4" tall, ivory or light brown donkey with painted vinyl rider and attached sombrero, Thomas Mfg. Corp., USA (No. 120), 1950-1951, Donkey Cart, donkey and cart without rider, (No. 76), 1949-1950, Plated Donkey (gold or silver) and Vinyl Rider, (No. 123), 1951, Plated Donkey only (No. 121), 1950.

Roman Chariot with Driver, boxed 7-1/2" x 2-5/8" x 3-3/4", gold chariot and black horses with detachable silver vinyl driver, plastic wheels, Thomas Mfg. Corp., USA (No. 337), 1956 to late 1950s. Suggested Retail $0.49. Mounted Roman Centurion, boxed, (No. 302), 1955 to late 1950s. Suggested Retail $0.25. Roman Legion, boxed, includes; one chariot and driver, one mounted centurion, four legionnaires marching with spears and four legionnaires marching with swords, (No. 297), 1955 to late 1950s. Suggested Retail $1.98.

Farm Wagon, 10" x 3-1/2" x 5-1/2", with farmer, wind-up motor with attached key that produces a "true-to-life" farm horse gait, assorted color combinations, plastic wheels, Wolverine Supply and Mfg. Co., USA (No. 34), 1952 to mid 1950s. Suggested Retail $1.98. Not shown, Sulky Racer, (No. 23), 1952 to mid 1950s. Suggested Retail $1.98.

POTPOURRI 285

Coca Cola Bottle Vending Machine Bank, 5-1/2" x 2-1/2" x 2", when coin is inserted a 1-3/16" tall miniature bottle of coke is dispensed, comes with seven bottles, red with white painted details, 20th Century Products, USA, molded by Precision Plastics Co., USA (No. 333), 1950 to early 1950s.

Flower Cart with Scoop, Cart 3-3/4" x 2-1/2" x 1-1/4", assorted color combinations, plastic wheels, Manufacturer unknown, USA, early 1950s.

Snub Nosed .38 Revolver, 7" long, shoots harmless pellets, revolving cylinder, green and silver, Manufacturer unknown, USA, mid 1950s, very similar to Renwal's Military and Police .38 Automatic Revolver 1955 (No. 265), 1955, possibly made from Renwal mold.

Carnival Rides, Boats, 3-1/4" x 2-1/4" x 2-3/8", (No. 12), Jets, 3-1/4" x 1-1/4" x 3-3/4", (No. 13), Race Cars, 1-1/4" high and 3-3/4" diameter, (No. 11), assorted color combinations, MARVI, USA, early to mid 1950s.

BIBLIOGRAPHY

BOOKS

Clauser, Henry R., editor. *Encyclopedia/Handbook of Materials, Parts and Finishes*. Technomic Publishing Co., Inc.. Westport, CT: 1976.

DuBoise, J. Harry. *Plastics History U.S.A.* Cahners Books. Boston: 1972.

Dunham, Arthur. *Working With Plastics*. McGraw-Hill. New York: 1948.

Dyer, Davis and Sicilia, David B. *Labors of a Modern Hercules, The Evolution of a Chemical Company*. Harvard Business School Press. Massachusetts: 1990.

Freeman, Ruth and Larry. *Cavalcade of Toys*. Century House. New York: 1942.

Hertz, Louis H. *The Toy Collector*. Funk and Wagnalls. New York: 1969.

Hertz, Louis H. *The Complete Book of Building and Collecting Model Automobiles*. Crown Publishers, Inc. New York: 1970.

McClintock, Marshall and Inez. *Toys in America*. Public Affairs Press. Washington, D.C.: 1961.

O'Brien, Richard. *Collecting Toy Soldiers*. Books Americana Inc. Florence, Alabama: 1992.

ARTICLES

Parker, L. C. "Just For Fun". *Modern Plastics*. Dec. 1938 p. 29.

"Playthings". *Modern Plastics*. Dec. 1938 p. 42.

Lougse, E. F. "Toys For Sale". *Modern Plastics*. Aug. 1939 p. 21.

"Plastics for Playthings". *Playthings*. March 1941 p. 196.

"Toying With Nitrocellulose". *Modern Plastics*. April 1941 p. 45.

"Playing Up A Market". $ST2Modern Plastics$ST1. July 1941 p. 43.

Knowles, Eleanor N. "Plastics, Priorities and Playthings". *Playthings*. March 1942 p. 190.

"Realities or Reveries". *Modern Plastics*. April 1944 p. 73.

Auerbach, Alfred. "I Don't Know". *Modern Plastics*. Oct. 1945 p. 99.

"Time To Play". *Modern Plastics*. Dec. 1945 p. 95.

"Plastic Toys". *Modern Plastics*. June 1947 p. 165.

"Toys". *Modern Plastics*. May 1948 p. 75.

"What Man Has Joined Together". *Fortune*. March 1936 p. 69.

"Plastics in 1940". *Fortune*. Oct. 1940 p. 89.

"Louis Marx: Toy King". *Fortune*. Jan. 1946 p. 122.

Glows-Mobile, 5 3/8" x 2" x 1 3/4", when front end is pushed down, metal strip makes contact with bulb in center of grill causing it to light up, comes with one AA battery, blue, plastic wheels, Great American Plastics Co., USA, late 1940s.

PRICE GUIDE

The values below are for toys in new or like-new condition. Since many plastic toys came from the factory with slight scratches or scuff marks, these defects are acceptable as long as they do not adversely affect the toys paint, decals, vacuum metalized finish, or hot stamping, where applicable. Deep scratches, cracks, chips, melt marks, missing pieces or accessories, repainting, or non-original type wheels greatly reduce the value of a toy. The same applies to toys that have friction or wind-up motors that no longer work.

A toy molded of cellulose acetate or cellulose acetate butyrate can be expected to exhibit a slight amount of warping. This should not affect its value, unless certain articulated parts such as doors, hood or trunk lid no longer open and close. When warping is so severe that a toy's friction or wind-up motor no longer functions or the wheels no longer turn, a price adjustment is in order.

When one or more of the above defects are encountered, it is up to the buyer and seller to negotiate a price acceptable to both parties.

A new or like new toy, still in its original box, is worth an additional 20-30%, as long as the box is complete and in excellent or better condition.

In most cases, the molds used to produce the toys in this book were sold for scrap long ago. However, some have survived. A few of these have been used to make reproductions. As of this writing, the Marx "Yellow Cab" on page 51, the Archer spacemen and women on page 195 and the Marx futuristic vehicles on pages 204 and 205 have all been reproduced. While a knowledgeable collector should be able to tell the difference, a novice may not be so lucky.

Other considerations affecting value are economic conditions, trends, and geographic location. Toys encountered at general antique shows and sales will usually be priced lower than those at antique toy shows, and being at the right place at the right time can make all the difference!

The lefthand number is the page number. The letters following it indicate the position of the photograph on the page: t=top, l=left, tl=top left, tr=top right, c=center, cl=center left, cr=center right, r=right, b=bottom, bl=bottom left, br=bottom right. The number in the center column is the estimated retail price range (in U.S. dollars) for a toy in new or like new condition. In photos with more than one object, individual items are identified by a key word or words on the right.

page	price ($)	item	page	price ($)	item	page	price ($)	item	page	price ($)	item
1	100-125		33b	5-10		42t	10-15	each	50cl	30-35	
2	100-125		34tl	75-100		42cr	30-35		50cr	20-25	
3	175-200		34tr	20-25		42cl	100-125		50b	45-50	each
4	45-50		34c	125-150		42b	100-125		51t	20-25	
6	15-20		34b	15-20		43t	100-125		51cr	100-125	
8	45-50		35t	15-20		43c	150-175		51b	30-35	
9	10-15		35cr	15-20		44t	125-150		52tl	75-100	
18	20-25		35cl	40-45		44c	125-150		52tr	45-50	
20	25-30		35b	30-45		44b	65-75		52cl	35-40	
22	150-175	motor bike	36tl	10-15		45t	175-200		52cr	20-25	
	55-60	station wagon	36tr	10-15	each	45cl	75-100		52b	30-35	
	55-60	convertible	36b	10-15	each	45cr	65-75		53t	100-125	
			37t	40-35		45b	65-75		53c	150-175	
	65-75	sedan	37cr	30-35		46t	75-100		53b	225-250	
26	50-75		37cl	35-40		46cr	100-125		54t	10-15	each
27	25-30		37b	20-25		46cl	65-75		54cr	30-35	
28	75-100		38t	20-25	sedan	46b	125-150		54cl	15-20	
30	30-35	sedan		15-20	trailer	47t	100-125		54b	15-20	
	35-40	truck	38c	20-25	coupe	47cl	250-300		55t	30-35	
	30-35	coupe		15-20	trailer	47cr	250-300		55cr	35-40	
	40-45	bus	38b	125-150	with wind-up	47b	45-50		55cl	35-40	
	30-35	taxi		100-125	without wind-up	48t	150-175		55b	60-65	
	65-75	DC-3				48cr	30-35		56t	55-60	
31t	15-20		39t	100-125		48cl	20-25		56cr	65-75	
31bl	30-35	each	39c	35-40		48b	100-125		56cl	55-60	
32t	25-30		39b	100-125		49t	30-35		56b	70-75	
32c	5-10		40t	75-100		49cl	30-35		57t	30-35	in hard plastic only
32b	10-15	each	40cr	45-50		49cr	30-35				
33tl	25-30		40cl	45-50		49b	30-35		57cl	150-175	
33tr	10-15		40b	100-125		50tl	30-35		57cr	30-35	
33c	25-30		41tr	125-150		50tr	20-25		57b	35-40	

58t	25-30		78t	20-25			15-20	farm truck	116t	250-275	
58cr	25-30		78cr	25-30			15-20	pickup truck	116cl	150-175	
58cl	45-50		78b	25-30	puzzle		15-20	oil truck	116cr	85-100	
58b	125-150			45-50	whistle	98t	20-25	each	116b	125-150	
59t	45-50		79t	30-35		98cr	45-50		117t	100-125	
59cr	45-50		79cl	55-60		98cl	15-20		117c	125-150	
59cl	60-65		79cr	25-30		98b	65-75		117b	175-200	
60t	100-125		79b	150-175		99t	10-15	each	118t	250-275	
60cl	20-25	coupe	80t	275-300		99cl	35-45	each with	118c	125-150	
	35-40	fire chief	80cr	75-100				load	118b	45-50	
	35-40	police	80cl	75-80			15-20	boat alone	119t	25-30	
60cr	20-25		80b	20-25		99cr	25-30	each	119cl	25-30	panel truck
60b	25-30		81t	175-200		99b	15-20	each	119cl	45-50	bus
61t	150-175		81cr	20-25		100t	20-25	each	119b	35-40	
61cl	40-45		81cl	15-20		100cr	25-30		120t	40-45	
61cr	10-15	each	81b	20-25		100cl	30-35		120cr	20-25	
61b	50-55		82t	200-225		100b	45-50	ladder truck	120cl	20-25	
62t	60-65		82cl	125-150			35-40	pumper truck	120b	20-25	each
62cr	200-250		82b	75-100		101t	50-60		121tl	25-35	
62cl	100-125		83t	150-175		101cl	45-50		121tr	75-100	
62b	200-250		83b	150-175		101cr	25-30		121c	30-35	
63t	10-15		84t	150-175		101b	15-20	each	121b	150-175	
63cl	25-30		84cr	75-100		102t	15-20	short truck	122t	25-30	each
63cr	25-30		84cl	20-25			25-30	long truck	122cr	150-175	
63b	75-100		84b	45-50		102cl	15-20		122cl	150-175	
64t	65-75		85t	20-25		102cr	15-20		122b	150-175	
64cr	20-25		85cl	20-25		102b	15-20		123t	75-100	
64cl	15-20	convertible	85cr	20-25		103t	20-25		123cr	100-125	
	30-35	trailer	85b	225-250		103cr	20-25		124t	100-125	
64b	20-25	limousine	86tl	25-30	without	103cl	20-25		124cr	35-40	
	25-30	trailer			driver	103b	3-5	each	124cl	25-30	
65t	15-20	sedan		20-25	with driver		15-20	station	124b	35-40	
	45-50	sedan with	86c	20-25				wagon	125t	35-40	
		gas tank	86b	100-125		104t	15-20	wrecker	125clt	20-25	each
65cl	30-35	sedan with	87t	100-125			20-25	service truck	125clb	10-15	each
		rack and	87cr	45-50		104cr	30-45	highway fleet	125b	20-25	
		canoe	87cl	150-175	without	104cr	35-40	pumper truck	126t	15-20	
65cr	25-30	police			motor	104b	20-25	car	126cr	15-20	
	25-30	fire chief		175-200	with motor		35-40	ladder truck	126cl	125-150	
65b	15-20	sedan	87b	25-30		105t	35-40		126b	30-35	
	45-50	trailer	88t	75-100		105cr	35-40		127tl	35-40	
66t	15-20	convertible	88cr	150-175		105cl	35-40		127tr	100-125	
66cl	30-35		88cl	100-125		105b	35-40		127b	100-125	
66cr	175-200	with motor	88b	10-15	each	106t	30-35	truck	128tr	40-45	
	125-150	without	89t	20-25			15-20	trailer	128tl	20-25	
		motor	89cr	55-65		106cr	20-25		128cr	75-100	
66b	75-100		89cl	25-30		106cl	20-25		128cl	30-35	
67t	55-65		89b	40-45		106b	25-30		128b	35-40	
67cr	40-45		90t	100-125		107t	20-25		129t	35-40	each
67cl	40-45		90c	25-30	race car only	107cl	35-40	each	129cr	35-35	
67b	40-45			40-50	race car and	107cr	35-40		129cl	45-50	
68t	25-30				truck	107b	45-50		129b	45-50	
68cl	25-30		90b	20-25		108t	45-50		130t	125-150	
68cr	15-20		91tl	75-100		108cl	45-50		130cr	20-25	
68b	40-45		91tr	175-200		108cr	65-75		130cl	80-100	
69t	20-25	Stutz	91b	175-200		108b	45-50		130b	45-50	
	20-25	roadster	91b	20-25		109t	45-50		131t	45-50	
	30-35	touring	92	35-40		109cr	55-65		131cr	100-125	
69b	45-50		93t	30-35		109cl	40-45		131cl	40-45	each
70	30-35		93b	15-20	each	109b	35-40		131b	45-50	
71	35-40		94t	30-35	mixer	110t	35-40		132t	45-50	
72tl	25-30			10-15	wheelbarrow	110cr	65-75	each	132cr	40-45	
72tr	35-40		94cl	30-35		110cr	30-35		132cl	20-25	
72c	20-25		94cr	20-25		110b	25-30		132b	20-25	
72b	20-25		94b	30-35		111t	25-30		133t	20-25	
73t	35-40		95t	65-75		111cr	25-30		133cl	225-250	
73cr	25-30		95cr	100-125		111cl	45-50		133cr	175-200	
73cl	20-25		95cl	45-50		111b	40-45		133b	200-225	
73b	20-25		95b	35-40		112t	75-100		134t	100-125	each
74t	40-45		96t	50-55	convertible	112c	85-100		134cl	25-30	
74b	100-125			50-55	fire truck	112b	225-260		134cr	25-30	
75t	65-75		96cl	45-50		113c	45-50		134b	25-30	
75c	25-30	with trailer	96cr	20-25		113b	45-50		135t	10-15	each
	15-20	without	96b	20-25		114t	75-100		135cr	250-300	
		trailer	97t	10-15	each	114cr	30-35		135cl	250-300	
75b	25-30		97cr	15-20		114cl	25-30		135b	150-175	
77t	20-25		97cl	15-20	plastic	114b	15-20	each	136t	175-200	
77cl	10-15	each			chassis	115t	175-200		136cr	65-75	long truck
77bl	45-50			25-30	tin chassis	115c	125-150			35-45	short truck
77br	35-40		97b	25-30	dump truck	115b	75-100		136cl	30-35	

136b	375-400		156c	175-200		178b	75-100		201t	25-30	car
137t	75-100		157t	125-150		179tl	20-25			15-20	gun
137c	75-100		157c	125-150	each	179tr	20-25	each	201cr	275-300	
137bl	15-20		157b	65-75		179c	75-100	each	201cl	125-150	
137br	20-25		158t	65-75		179b	150-175		201b	75-100	
138t	35-40		158cr	20-25	cabin cruiser	180t	30-35		202t	125-150	
138cr	35-40		158cl	45-50		180cr	35-40		202cl	45-50	each
138cl	35-40	early version	158cr	5-10	small boats	180c	3-5	each	202cr	100-125	
	20-25	late version	158b	55-60		180b	75-100		202b	75-100	
138b	25-30		159tl	20-25		181t	75-100		203t	100-125	
139t	25-30		159tr	10-15		181c	5-10	each	203cr	225-250	
139cl	15-20	delivery truck	159c	35-40		181b	30-35		203cl	15-20	each
	20-25	sound truck	159b	15-20	each pirate	182t	100-125		203b	15-20	spacemen
	25-30	repair truck	160t	30-35		182cr	30-35			3-5	soldiers
	20-25	tow truck	160c	30-35		182cl	30-35		204t	65-75	
139cr	35-40		160b	35-40		182b	30-35		204cr	55-60	small
139b	40-50	truck and race car	161t	300-340		183t	20-25			40-45	large
	25-30	race car only	161cr	75-85	destroyer	183cr	20-25		204cl	40-45	
			161cl	75-85		183cl	20-25		204b	40-45	
140tl	30-35		161cr	75-85	aircraft carrier	183b	65-75		205t	40-45	
140tr	30-35					184t	25-30		205cl	65-75	
140c	35-40		161b	75-85		184c	25-30		205cr	65-75	
140b	20-25		162t	125-150		184bl	5-10		205b	100-125	
141t	30-35		162cr	120-155		184br	30-35		206t	8-10	each
141cr	40-45		162cl	20-25		185t	25-30		206cr	15-20	small
141cl	25-30		162b	20-25		185c	75-100			20-25	medium
141b	20-25		163t	15-20		185b	30-35			50-75	large
142t	25-30		163cr	20-25	freighter	186t	45-55		206cl	50-75	
142cr	30-35		163cl	15-20		186c	30-35		206b	15-20	small
142cl	15-20		163cr	15-20	motor cruiser	186b	30-35			20-25	medium
142b	10-15		163b	15-20		187t	30-35		207t	130-175	
143	45-50		164t	75-85		187c	40-45		207cl	35-40	
144	15-20		164cr	65-75		187b	200-235		207cr	45-50	
145	20-25	each	164cl	55-60		188t	30-35	without peg	207b	100-125	
146	35-40	large submarine	164b	100-125		188cr	30-35	without peg	208t	175-200	
			165t	200-250		188cl	15-20		208cl	65-75	
	15-20	small submarine	165b	65-75		188b	15-20		208cr	35-40	
			166t	15-20	each	189t	100-125		208cl	65-75	
	45-50	large frogmen set	166cl	75-100		189cl	20-25		208b	40-45	
			166cr	35-40		189cr	35-40		209t	200-225	
	15-20	small frogman	166b	20-25		189b	20-25		209cl	30-35	each
			167t	5-10		190t	25-30		209cr	225-250	
	45-50	large PT boat	167c	30-35		190c	45-50		209b	25-30	
	15-20	small PT boat	167b	145-175		190b	25-30		210	45-50	wrecker
			168t	25-30		191t	55-65			20-25	staff car
147t	100-125		168cr	60-80		191cl	45-50		211t	20-25	
147cr	45-50		168cl	25-30		191cr	25-30		211c	25-30	each
147cl	15-20		168b	30-35		191b	25-30		211b	75-100	
147b	15-20	carrier	169t	35-40		192t	20-25	B-17	212t	5-10	each
	10-15	battleship	169cl	35-40			20-25	B-36	212cl	15-20	
148t	25-30		169cr	30-35			20-25	set	212cr	35-40	
148cr	15-20	submarine	169b	25-30		192bl	10-15		212b	15-20	
148cl	20-25	PT boat	170t	10-15		192br	10-15		213t	15-20	each
148cr	5-10	pleasure boats each	170cr	10-15		193	235-275		213cl	30-35	
			170cl	30-35		194t	150-175		213cr	35-40	
148b	20-25		170	15-20		194b	150-175		213b	40-45	
149tr	45-50		171t	65-75		195t	45-50	women	214t	15-20	
149tl	20-25		171b	55-65			30-35	boy	214cl	35-40	lift truck
149cr	20-25		172	55-65	Dillon Beck	195b	8-10	men	214cr	40-45	handcar
149cl	100-125			30-35	Ideal		20-25	robot	214bl	15-20	
149b	175-200		172b	20-25	A-20	196t	8-10	each	214br	20-25	
150t	20-25			20-25	Defiant	196cl	15-20	men	215tl	20-25	
150cl	20-25			35-40	P-40		35-40	robot	215tr	20-25	
150cr	150-175		173	25-30		196cr	45-50	each	215cl	25-30	
150b	35-40		174	20-25		196b	75-85		215cr	20-25	
151	225-250		175t	10-15		197t	200-250		215b	25-30	
152t	75-100		175cr	30-35		197cr	30-35		216t	30-35	
152cr	100-125		175cl	35-40		197cl	75-100		216cl	50-55	
152cl	20-25	each	175b	35-40		197b	30-35		216cr	50-55	
152b	100-125		176tl	35-40		198t	75-100		216b	40-45	
153t	100-125		176tr	35-40		198cr	30-35		217t	40-45	
153c	75-100		176c	35-40		198cl	75-100		217cr	30-35	
153b	75-100		176b	75-100		198b	30-35		217cl	65-75	
154t	100-125		177tl	75-100		199t	75-100		217b	65-75	
154cr	300-325		177tr	75-100		199cl	30-35		218t	20-25	
154b	75-100		177c	100-125		199cr	75-100		218cr	75-100	
155t	65-75		177b	65-75		199b	200-250		218cl	25-30	
155c	150-175		178tl	20-25		200t	35-40		218bl	35-40	
155b	100-125		178tr	20-25		200c	125-150		218br	30-35	
156t	150-175		178c	65-75		200b	25-50	each	219t	30-35	

Code	Price	Description	Code	Price	Description	Code	Price	Description	Code	Price	Description
219c	10-15	each		50-55	lawn mower		30-35	chest		15-20	lamp
219b	50-75		241cr	65-75		251cr	35-40	buffet		10-15	club chair
220t	65-75		241cl	65-75			25-30	table	258cl	15-20	console radio
220c	30-35		241b	65-85	set		10-15	chair		25-30	fireplace
220b	25-30			15-20	blue sink		35-40	china closet		25-30	radio
221t	75-100				(No. HBB10)	251cl	35-40	refrigerator			phonograph
221cl	75-100			15-20	blue water		30-35	table	258cr	25-30	piano
221cr	55-65				closet (No.		10-15	chair		10-15	bench
221b	25-30				HBWC25)		35-40	range		10-15	lamp
222t	35-45			15-20	blue hamper		35-40	sink		15-20	table
222cr	50-55	without gun			(No. HBH10)	251b	20-25	each chair	258b	180-245	living room
222cl	35-45	radar truck		15-20	blue bathtub			section			set
222cr	35-45	searchlight			(No. HBT25)		20-25	coffee table	259cr	15-20	rocking chair
		truck	242t	10-15	each		25-30	lamp		20-25	baby bath
222b	65-75		242c	30-35	each	252tl	100-125	television		15-20	highboy
223t	15-20		242br	20-25			50-75	man	259cl	20-25	cradle
223cl	15-20		243t	10-15	each		50-75	woman		20-25	high chair
223cr	15-20		243c	75-100	bridge set		50-75	boy		25-30	play pen
223b	30-35		243b	25-30	breakfront		50-75	girl	259b	25-30	each
224t	45-50			25-30	buffet	252tr	10-15		260t	15-20	tricycle
224c	180-210			20-25	arm chair	252cr	20-25			35-40	swing
224bl	45-50			15-20	side chair	252b	30-35	push cart		25-30	see-saw
224br	35-40			25-30	table		20-25	doll on toilet		25-30	slide
225t	30-35		244cr	230-275	garden set	253t	20-25	jukebox		30-35	kiddie car
225cr	25-30		244cl	10-15	lawn chair		8-10	sister	260cl	20-25	student's
225cl	25-30			10-15	lawn bench		8-10	brother			desk
225cr	30-35			45-50	trellis		20-25	bar		35-40	teacher's desk
225b	30-35			45-50	pool	253cl	5-10	each		25-30	teacher's
226t	30-35	each		10-15	lounge chair	253b	25-30				chair
226cr	190-250		244bl	20-25	each	254t	5-10	each	260cr	35-40	
226cl	25-30		244br	25-30	lawn table	254cr	5-10	each	260b	15-25	each
226b	15-20			45-50	picnic table		8-10	mother	261t	30-35	
227t	15-20		245t	15-20	each		8-10	sister	261cl	15-20	
227cr	15-20	stake trailer		5-10	chair	254cl	5-10		261cr	15-20	small dolls
227cl	15-20		245cr	55-65	cabinet	254b	15-20	telephone		30-35	"Chubby"
227cr	20-25	with load		30-35	refrigerator		10-15	radio			doll
227cl	15-20			35-40	range		10-15	bath scale	261b	90-100	doll family
227b	15-20			45-50	sink		25-30	baby scale			set
228t			245cl	45-50			10-15	stool	262t	20-25	mother
228cr	30-35	long-bed	245b	35-40	washer		20-25	smoking		20-25	brother
		wrecker		30-35	ironer			stand		45-50	doctor
228cl	65-75		246t	60-90	living room		10-15	mantle clock		45-50	nurse
228b	20-25	each			set		15-20	step-on can		20-25	father
229t	30-35		246cr	10-15	each		10-15	dust pan		20-25	sister
229cr	25-30		246cl	10-15	each		10-15	kitchen clock		55-60	policeman
229cl	25-30	tow truck	246b	20-25	radio	255t	10-15	each		55-60	mechanic
	20-25	staff car		35-40	fireplace	255cr	15-20	server	262cl	5-8	each
229b	35-40	road roller		20-25	radiator		15-20	table	262cr	15-20	each
	40-45	compressor	247t	75-100			5-10	chair	262cl	20-25	sweeper
230t	10-15		247cr	20-25			15-20	china closet	262b	50-75	triple swing
230c	15-20		247cl	50-55	piano and		15-20	buffet		20-25	slide
230b	20-25				bench		15-20	highboy		35-40	hammock
231t	30-35		247b	40-45		255cl	5-10	night table		25-30	single swing
231cr	10-15		248t	75-100			15-20	bed		50-75	see-saw
231cl	20-25		248cl	105-135	nursery set		15-20	vanity			swing
231b	35-40		248b	25-30	crib		15-20	dresser		75-100	ferris wheel
232t	75-100			45-50	playpen	255bl	100-125	bridge set		35-45	duo-swing
232cr	30-35			30-35	high chair	255br	20-25	vacuum	263t	20-25	sled
232cl	35-40		249t	35-40	stroller		35-40	sweeper		3-5	dog
232b	35-40			30-35	carriage		25-30	mop		3-5	cat
233t	45-50		249cl	20-25	chair		25-30	broom	263cl	20-25	
233b	20-25			25-30	cradle	256t	90-135	dining room	263cr	35-40	
234t	15-20		249cr	20-25				set	263b	5-10	each
234cr	65-75	cable trolley	249b	30-35		256cr	80-120	kitchen set	266t	25-30	
234cr	45-50	tricky trolley	250tr	120-145	bedroom set	256cl	15-20	stove	266b	20-25	
234b	150-175		250cl	15-20	"Boopsie"		15-20	refrigerator	267t	20-25	each
238	15-20	street light			doll		15-20	sink	267cl	5-10	each
	10-15	watering can		20-25	standing doll		15-20	table	267cl	5-8	each
	20-25	lawn mower		25-30	sunsuit doll		5-10	chair	267b	5-8	each
	25-30	bird cage		20-25	jointed doll	256b	20-25	stove	268t	8-10	each
	10-15	wheelbarrow	250cr	35-40	bathtub		20-25	refrigerator	268cl	45-55	dairy set
240t	20-25	sink		15-20	chair		20-25	sink	268cr	65-75	circus set
	10-15	girl		35-40	washstand		15-20	table		20-25	girl on
240cl	30-35			20-25	water closet		5-10	chair			elephant
240cr	55-65			20-25	waste can	257t	75-100	sink	268b	65-75	infantry set
240b	45-50		250b	20-25			75-100	refrigerator	269t	175-200	cowboy and
241t	20-25	vacuum	251t	30-35	dresser		75-100	stove			Indian set
		cleaner		15-20	bench	258t	10-15	barrel chair	269cr	175-200	infantry and
	50-55	carpet		10-15	night table		10-15	sofa			cavalry set
		sweeper		35-40	bed		10-15	table	269cl	25-30	

269b	20-25			273t	150-175		278t	75-100		282cr	175-200	
270t	25-30			273cr	30-35		278cr	25-30		282b	75-100	
270cl	25-30	egg		273cl	75-100		278cl	125-150		283t	175-200	
	35-40	cart		273b	30-35		278b	55-75		283cl	45-50	zoo
	20-25	basket		274t	100-125		279tl	100-125		283cl	125-150	bridge
270cr	35-40			274cl	50-75	three blind mice	279tr	100-125		283b	15-20	
270b	15-20	"Suzie"					279c	55-65		284t	175-200	
	20-25	shaver		274cr	100-125		279bl	45-50		284cl	100-125	superette
	20-25	telephone		274b	100-125	each	279br	20-25		284cl	15-20	each
	15-20	dog		275t	75-100		280t	10-15	each	284b	20-25	each
	15-20	"Suck-a-Thumb"		275cl	65-75		280cr	45-50		285t	20-25	(No. 120)
				275cr	150-175		280cl	100-125			35-40	(No. 76)
271t	50-75			275b	15-20		280b	25-30		285c	109-135	(No. 297)
271cl	50-75			276t	125-150		281t	65-75			10-15	(No. 302)
271cr	150-175			276c	175-200		281cr	75-100	covered wagon only		35-40	(No. 337)
271bl	100-125			276b	125-150					285b	65-75	each
271br	65-75			277t	200-225		281cl	45-55	buckboard only	286t	125-150	
272t	200-225	early		277c	100-125					286cr	15-20	
	150-175	mid		277bl	150-175		281b	20-25	each	286cl	20-25	each
	125-150	late		277br	75-100	(No. 4577)	282t	125-150		286b	25-30	each
272c	125-150				125-150	(No. 4578)	282cl	100-125		287	75-100	
					150-175	(No. 4579)						

MANUFACTURERS AND DISTRIBUTERS

Ajax Plastics, New York, NY
Ardee Plastics Co., Inc., Long Island City, NY
Argo, (city unknown)
All Metal Products Co. (Wyandotte Toys), Wyandotte, MI
Alladin Plastics, Inc., Los Angeles, CA
Allied Molding Corp., Flushing, NY
Allstate Engineering Service, Dayton, OH
Amerline, Inc., Chicago, IL
Archer Plastics, Inc., New York, NY

Bachmann Brothers, Inc., Philadelphia, PA
Banner Plastics Corp., New York, NY
Beemak Plastics, Los Angeles, CA
Bell Products Co., St. Louis, MO
Bergen Toy and Novelty Co., Carlstadt, NJ
Best Plastics Corp., Brooklyn, NY
B. W. Molded Plastics, Inc., Pasadena, CA

Caldwell Products, Inc., Bronx, NY
California Moulders, Inc., Los Angeles, CA
Clinford Corp., West New York, NJ
Commonwealth Plastics Corp., Leominster, MA
Conway Co., Skokie, IL
Cruver Mfg. Co., Chicago, IL

Deluxe Game Corp., Richmond Hill, Long Island, NY
Dillon Beck Mfg. Co. (Wannatoy), Hillside, NJ

Elmar Plastic Corp., New York, NY
Empire Plastic Corp., Bronx, NY

Gilmark Merchandise Corp., New York, NY
Gerber Plastics Co., St. Louis, MO
Great American Plastics Co., Fitchburg, MA

Hardy Plastics and Chemical Corp., Brooklyn, NY
Hassenfeld Brothers Co. (Hasbro), Central Falls, RI
Hubley Mfg. Co., Lancaster, PA

Ideal Toy Corp. (Ideal Novelty and Toy Co.), Hollis, Long Island, NY
Irwin Corp., Fitchburg, MA

K & O Models, Inc., Van Nuys, CA
KAY-DEE Plastics Inc., New York, NY
Keystone Mfg. Co., Boston, MA
Kiddie Brush and Toy Co., Jonesville, MI

Kilgore Mfg. Co., Westerville, OH
Knickerbocker Plastic Co., Inc., Glendale, CA

Lapin Products Co., Newark, NJ
Lido Toy Corp., New York, NY
Lincoln Line, Inc., Chicago, IL
Louis Marx and Co., New York, NY

Mattel Creations, Culver City, CA
Manoil Co., Waverly, NJ
Modern Plastics Corp., Benton Harbor, MI

Nosco Plastics (National Organ Supply Co.), Erie, PA

Palmer Plastics Mfg., Inc., Woodside, Long Island, NY
Plas-Tex Corp., Los Angeles, CA
Plastic Art Toy Corp. of America, East Rutherford, NJ
Plastic Toys, Inc., Byesville, OH
Plasticraft Mfg. Co., Arlington, NJ
Product Miniature Co., Inc., Milwaukee, WI
Precision Plastics Co. (Plastic Masters, Triumph, U.S. Plastics), Philadelphia, PA
Premier Products Co., Brooklyn, NY
Pressman Toy Corp., New York, NY
PRE-VIEW Toy and Novelty Co., Inc., N.Y., NY
Processed Plastics Co., Aurora, IL

Renwal Mfg. Co., Inc., Mineola, NY
Ross Tool and Mfg. Co., New York, NY
Royal Plastics, Inc. (city unknown)
Revell Plastics, Los Angeles, CA

Sidney A. Tarrson Co. (Beco Plastics Co.), Chicago, IL
Saunders Tool & Die Co. (Saunders-Swadar Toy Co.), Aurora, IL
Slik-Toy, Lansing, MI

T. F. Butterfield, Inc. (Bonnie Bilt), Naugatuck, CT
Tedsco Plastics, Inc., Phoenix, AZ
Tee-Vee Toy, Inc. (Bradford Novelty Co., Inc.), New York, NY
Thomas Mfg. Corp. (Acme Plastic Toys, Inc.), Newark, NJ
Timely Toys, Inc., St. Louis, MO
Tom and Co., San Francisco, CA
Topic Toys, Chicago, IL
Toy Founders, Inc., Detroit, MI
Twentieth Century Products, New York, NY

INDEX

Acme Plastics Mfg. Co., Inc., 21, 26, 27, 71, 145, 172, 173, 189, 210, *see also* Thomas Mfg. Corp.
Acrylic, 5
Airplanes, 10, 30, 145, 171-192, 210
Ajax Plastics Corp., 212
All Metal Products Co., 13, 32, 94, 240, 266
Alladin Plastics, Inc., 12, 13
Allied Molding Corp., 69, 93-96, 240, 263
Allstate Engineering Service, 147
American Flyer *see* A. C. Gilbert Co.
Amerline, Inc., 147, 266
Ambulances, 33, 37, 40, 42, 57, 216, 219, 223
Appleby, Scott, 194, 195
Archer Plastics, Inc., 7, 96, 193, 195-199
Ardee Plastics Co., Inc., 230
Argo, 174
Automobiles, 22, 23, 30-69, 197, 198, 200, 202, 204, 205, 210-212, 216, 223, 224, 226, 229, 272, 273, 287

B. W. Molded Plastics, 13, 32, 77, 96, 230
Bachmann Brothers, Inc., 32, 33, 97,
Bakelite, 6
Bakelite Corp., 30
Banner Plastics Corp., 12, 13, 31, 33, 34, 92, 93, 97-100, 147, 210, 212-214, 231, 267
Beck, George, 143
Beemak Plastics, 195
Beetle, 6
Bell Products Co., 18
Bergman, David, 92
Bergman, Robert, 92
Bergman, Ross, 92
Bergen Toy and Novelty Co., 16, 28, 264-269
Berlin, Frank, 194, 195
Best Plastics Corp., 148
Beton, *see* Bergen Toy and Novelty Co.
Blow molding, 143

Bonnie Bilt, 211, 214
Boats, 143-170, 210, 217, 218, 224
"Bugle Boy," 9
Buses, 30, 31, 92, 97, 103, 115, 118, 119, 120, 126, 128, 134, 140

Caldwell Products, 263
California Moulders, Inc., 71, 72, 93, 100, 101, 121, 193, 194
Canada, 3, 57, 130, 145, 174
Carpenter, W. B., 5
Cassidy, Hopalong, 271
Celanese Corp. of America, 5, 8, 11, 15, 16, 17, 29
Celluloid, 5, 27, 143
Celluloid Mfg. Co., 5
Cellulose acetate, 5-7, 9, 10, 17, 20, 146
Cellulose acetate butyrate, 5, 8, 9
Cellulose nitrate, 5, 143
Character Molding Corp, 72
Chicago Musical Instrument Co., 9
Clinford Corp., 34
Coca Cola, 121, 286
Commonwealth Plastics Corp., 77, 94, 101, 210, 238, 231, 239
Compression molding, 6, 7
Consolidated Molded Products Corp., 26, 27, 29, 145, 172
Construction vehicles, 92-95, 98-103, 105, 107, 108, 110, 113, 114, 116, 121, 126, 128-135, 137, 140, 212, 213, 216, 229, 229
Convertibles, 8, 22-24, 32, 34, 42-48, 50, 53, 54, 56, 58, 60-64, 66, 68, 69, 96, 197, 204, 229
Conway Co., 142
Crockett, Davy, 275, 279, 281

Danien, Bud, 23
Danien, Edward W., 22, 23
Deluxe Game Corp., 31
Detroit Mold Engineering Co., 28

Dillon Beck Mfg. Co., 16, 29, 31, 34, 35, 70, 72, 77, 92, 93, 101-103, 143, 143-146, 148, 171-173, 175, 193, 200, 214
Dillon, Daniel C. Jr., 143, 144
Dimensional stability, 8
Doll house furniture, 95, 235-263
Doody, Howdy, 271, 284
Dow Chemical Co., 9, 11, 14, 15, 17, 29
"Dragnet," 273
Duncan, Ted, 264
Du Pont Co., 7

Elmar Products Co., 77, 231
England, 1, 5, 27, 29, 116, 231
Empire Plastic Corp., 78, 103, 231
Extrusion molding, 6

Fire trucks, 92-94, 96-98, 100-104, 112, 114, 116, 117, 129, 124, 127, 129, 132, 134-137, 140, 141, 208
Firestone Tire and Rubber Co., 29
Foster Grant Co., Inc., 200
Frutchey, Harold, 27

Gerber Plastic Co., 12, 13, 78, 240
Germany, 5, 6, 28, 143
Gilbert, A. C., Co., 12, 230
Gilmark Merchandise Corp., 35, 93, 104, 200, 201, 214, 215
Glen Dimension, 78
"Golden Books," 264, 274
Great American Plastics Co., 287

Hardy Plastics and Chemical Corp., 13, 263
Hasbro, Inc., 37
Hassenfeld Brothers Co., *see* Hasbro, Inc.
Helicopters, 21, 29, 173, 177, 183, 190, 191
Hirsch, Benjamin, 146
Hirsch, Henry, 146

Hirsch Laboratories, 146
Hot rods, 2, 55, 62, 76, 81, 83, 84
Hubley Mfg., Co., 37, 79, 104, 105, 174, 180, 215
Hyatt, John Wesley, 5

Ideal Novelty and Toy Co., see Ideal Toy Corp.
Ideal Toy Corp., 10-13, 16, 19, 20, 23-25, 27, 28, 31, 32, 37-47, 70, 72, 79, 80, 93, 106-118, 144-146, 148-157, 172-179, 193, 201-203, 210, 211, 216-219, 231, 232, 234, 235, 238-252, 264, 269-279
Injection molding, 7
Irwin Corp., 8, 12, 13, 37, 47, 48, 118, 119, 203, 252
Ivory, 48

Japan, 28, 32, 143, 144, 174, 266
Jeeps, 10, 21, 31, 42, 70-75, 179, 189, 210, 215, 216, 219, 220, 226, 227-229, 264, 276
Jericho Toy Mfg. Corp., 252
"Jewels For Playthings," 30, 92, 171

K & O Models, Inc., 158
Katz, Abraham, 29
Kay-Dee Plastics, Inc., 141
Keystone Mfg. Co., 32, 48, 49, 119
Kiddie Brush and Toy Co., 284
Kilgore Mfg. Co., 30, 31, 49, 92, 110, 119, 171
Kohner Brothers, 279
Korean War, 210, 211
Knickerbocker Plastics Co., 13, 158
Kleeware, 116, 231
Kugel, Charles, 76, 77
Kugel, Harry H., 76
Kugel, Ruben G., 76

Lapin Products Co., 31, 49, 50, 120
Lester, Betty, 210
Lester, William M., 193, 210
Lido Toy Corp., 50, 80, 120, 121, 141, 180, 203, 279, 280
Lincoln Line, Inc., 81
Lionel Mfg. Co., 230
Louis Marx & Co., 25, 29, 50 - 54, 69, 73, 74, 81, 121-125, 158, 159, 174, 180-182, 20-5, 210, 220-223, 239, 253, 254, 280, 281
Lumarith, 5, 8, 17, 30
Lupor Metal Products, Inc., 54
Lustrex, 11
Lustron, 11

Manoil Co., 54, 125, 126, 265
Marcak, Charles, 264
Marx, Louis, 26
Marx Toys, see Louis Marx and Co.
Mattel Inc., 13, 126, 205, 234
Mattel Creations, see Mattel Inc.
Melamine, 5
Melmac, 6
Mexico, 39, 72
Michtom, Morris, 27
Michtom, Benjamin F., 264
Military toys, 18, 20, 21, 143-192, 210-229
Modern Plastics Magazine, 9, 10, 23, 30, 31, 70, 171, 235, 265
Molders mark, 21, 25
Motorcycles, 3, 22-24, 76-80, 82, 84-91, 226
Monsanto Chemical Co., 9, 11, 12, 13, 15, 17
Multiple Products Corp., 159
Musical toys, 6, 8, 9

National Organ Supply Co., see Nosco Plastics
Neutrality Act, 144
Newark Die Co., 29
Nosco Plastics, 32, 55, 76, 77, 82, 83, 126, 127, 232, 233, 280
Nylon, 5

Parkes, Alexander, 5
Pepsi Cola, 271, 280
Phenolic plastics, 5
PLASCO, see Plastic Art Toy Corp. of America
Plas-Tex Corp., 70, 75, 128, 194, 195
Plastic Art Toy Corp. of America, 12, 13, 235, 239
Plastic Masters, see Precision Plastics Corp.
Plastic Merchandiser, 11, 15
Plastic Toys, Inc. , 145
Plastic toys
  care of, 18-20
  cleaning, 19
  history, 5-17
  identification, 21-25
  marketing, 10-17
  repair, 19
  value, 18
Plasticraft Mfg. Co., 128, 159, 205, 223, 281
Polyester, 5
Polyethylene, 5, 9

Polystyrene, 5, 9, 11, 17, 146
Precision Plastics Co., 13, 21-24, 32, 55, 56, 84
Premier Products Co., 56, 57, 183, 206, 207
Pressman Toy Corp., 57, 282
Pre-View Toy and Novelty Co., 183
Processed Plastics Co., 84, 92, 93, 128, 129, 223
Product Miniature Co., 130
Pyro Plastics Corp., 22, 84, 85, 93, 130, 160, 161, 184, 193, 207, 208, 211, 224-228

Quarnstrom, Ted, 28

Race cars, 42, 76-90, 92, 139, 209
Rel Plastics, 266
Reliable Plastics Co., 3, 57, 130, 131, 145, 174
Reliance Molded Plastics, Inc., 85
Remco Industries, 146
Renwal Mfg. Co., Inc., 12-14, 16, 23, 25, 32, 58-62, 86-88, 92, 93, 131-136, 162-165, 174, 184-188, 193, 209, 210, 234-238, 254-262, 282-284
Revell Plastics, 13, 29, 136, 137, 159
"Robert the Robot", 264, 272
Rogers, Roy, 70, 73, 264, 276, 277
Rosenfield, Max, 31
Ross Tool and Mfg. Co., 63, 88, 89, 137
Royal Plastics, Inc., 63
Rowan, Edward W., 29, 143-145

Saunders, Paul, 92
Saunders Tool & Die Co., 12, 13, 63, 69, 77, 89, 92, 165, 189, 228
Shapiro, Benjamin, 27, 145, 172, 173
Sharpie, O. J., 144, 145
Sidney A. Tarrson Co., 138
Slik-Toy, 64
Society of Plastic Engineers, Inc. (SPE), 10, 29
Society of the Plastics Industry, Inc. (SPI), 10
Space toys, 2, 7, 193-210
Sports cars, 35, 36, 44-46, 53, 54, 63
Station wagons, 22-24, 33, 34, 47, 51-53, 55, 56, 64, 103, 212
Sterling Plastics Co., 64
Styron, 9, 11, 14, 15, 29
Sun Dial Products, 263

Swader Toy Co., *see* Saunders Tool & Die Co.

T. Cohen, 141
T. F. Butterfield, Inc., *see* Bonnie Bilt
Talking toys, 265, 272-274
Taxis, 30, 39, 49, 51, 52, 59, 66
Tea sets, 27-29, 235, 252
Tedsco Plastics, Inc., 68
Tee-Vee Toys, Inc., 284
Teeny Toy Co., 262
Tenite, 5-8, 17, 143, 265
Thermoplastics, 5, 6
Thermosetting plastics, 5, 6
Tennessee Eastman Corp., 5, 8, 11, 15, 17, 143, 265
Thomas, Islyn, 21, 25-29, 70, 145, 172-174, 235
Thomas Mfg. Corp., 13, 17, 21, 25, 29, 31, 32, 64-66, 70, 71, 75, 76, 89-93, 138-140, 145, 146, 166-169, 172, 174, 189-191, 193, 210, 211, 228, 229, 234, 262-264, 285
Thomas Toy, *see* Thomas Mfg. Corp.
Topic Toys, 191
Toy Founders, Inc., 66
Tom and Co., 191
Tow trucks, *see* Wreckers
Tractors, 93, 99, 102, 107, 108, 109, 124, 136
Trailers, house and utility, 23, 32, 38, 39, 64, 65, 66, 93, 106, 211, 218

Trains, 230-233
Tri-Play Toys, Inc., 140
Triumph, *see* Precision Plastics Corp.
Trolleys, 103, 234
Trucks, 30, 92-142, 179, 198, 199, 202, 203, 208, 210, 213, 217-229
Twentieth Century Products, 170, 286

U. S. Plastics, Inc., *see* Precision Plastics Corp.
United States, 5, 6, 9, 32, 143, 144, 146

Vacuum metalizing, 19, 20
Variegated plastics, 18
"Victory Fleet", 28, 145
VIBRO-ROLL Products, 103, 192
Vinyl, 18, 19

Wannatoy, *see* Dillon Beck Mfg. Co.
Waterbury Companies, Inc., 6
Weintraub, Lionel, 29
Wolverine Supply and Mfg. Co., 235, 285
World War I, 143
World War II, 10, 11, 27, 28, 29, 31, 70, 144, 145
Wreckers, 40, 93, 103-105, 118, 123-126, 129, 138, 141, 167, 210, 219, 226, 228, 229
Wyandotte Toys, *see* All Metal Products Co.

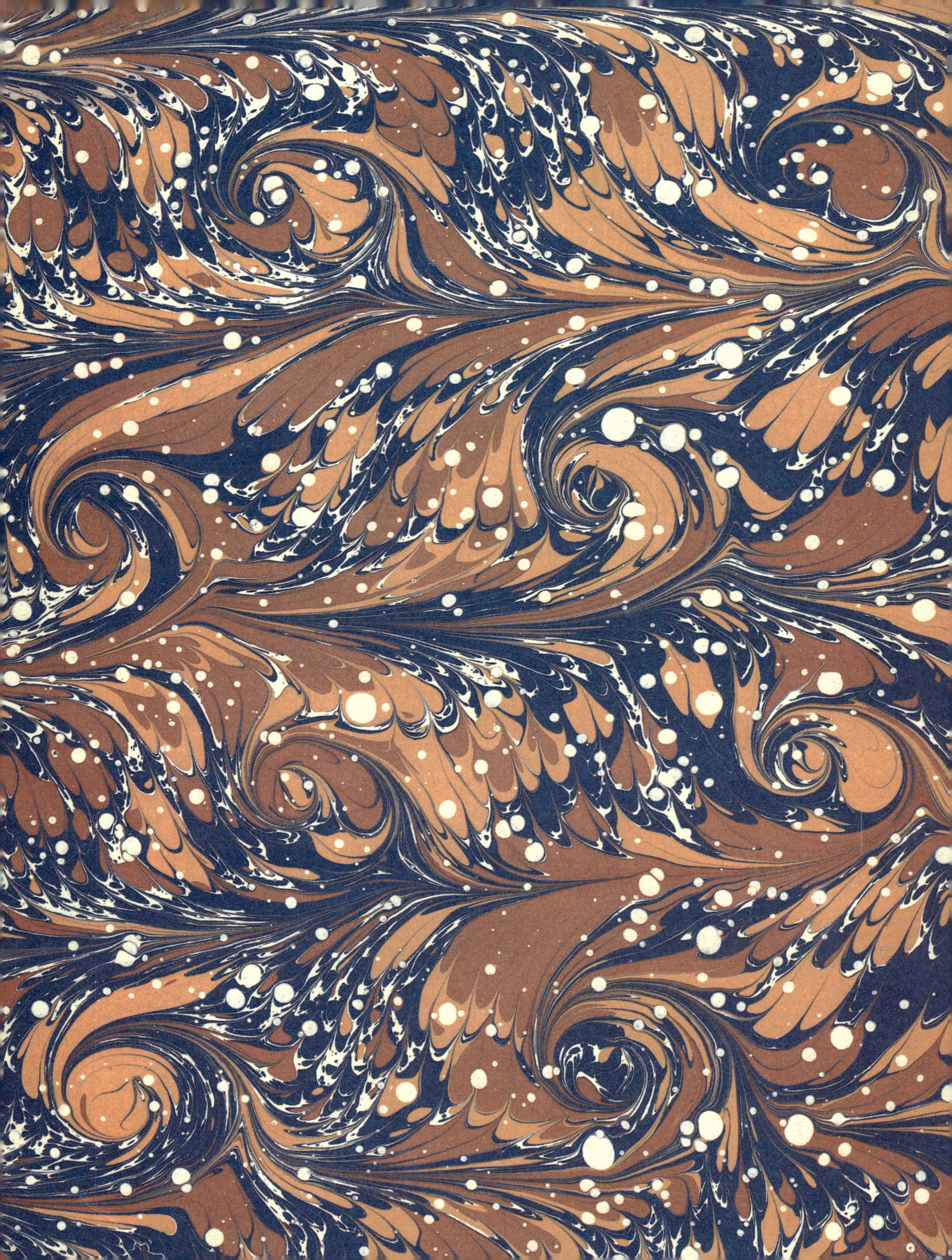